T0237285

Quantum Science and Technology

Series Editors

Raymond Laflamme, University of Waterloo, Waterloo, ON, Canada

Daniel Lidar, University of Southern California, Los Angeles, CA, USA

Arno Rauschenbeutel, Vienna University of Technology, Vienna, Austria

Renato Renner, Institut für Theoretische Physik, ETH Zürich, Zürich, Switzerland

Jingbo Wang, Department of Physics, University of Western Australia, Crawley, WA, Australia

Yaakov S. Weinstein, Quantum Information Science Group, The MITRE Corporation, Princeton, NJ, USA

H. M. Wiseman, Griffith University, Brisbane, QLD, Australia

Section Editor

Maximilian Schlosshauer, Department of Physics, University of Portland, Portland, OR, USA

The book series Quantum Science and Technology is dedicated to one of today's most active and rapidly expanding fields of research and development. In particular, the series will be a showcase for the growing number of experimental implementations and practical applications of quantum systems. These will include, but are not restricted to: quantum information processing, quantum computing, and quantum simulation; quantum communication and quantum cryptography; entanglement and other quantum resources; quantum interfaces and hybrid quantum systems; quantum memories and quantum repeaters; measurement-based quantum control and quantum feedback; quantum nanomechanics, quantum optomechanics and quantum transducers; quantum sensing and quantum metrology; as well as quantum effects in biology. Last but not least, the series will include books on the theoretical and mathematical questions relevant to designing and understanding these systems and devices, as well as foundational issues concerning the quantum phenomena themselves. Written and edited by leading experts, the treatments will be designed for graduate students and other researchers already working in, or intending to enter the field of quantum science and technology.

More information about this series at https://link.springer.com/bookseries/10039

Filip Wojcieszyn

Introduction to Quantum Computing with Q# and QDK

 Springer

Filip Wojcieszyn
Zürich, Switzerland

ISSN 2364-9054   ISSN 2364-9062 (electronic)
Quantum Science and Technology
ISBN 978-3-030-99381-8   ISBN 978-3-030-99379-5 (eBook)
https://doi.org/10.1007/978-3-030-99379-5

© The Editor(s) (if applicable) and The Author(s), under exclusive license to Springer Nature
Switzerland AG 2022
This work is subject to copyright. All rights are solely and exclusively licensed by the Publisher, whether
the whole or part of the material is concerned, specifically the rights of translation, reprinting, reuse
of illustrations, recitation, broadcasting, reproduction on microfilms or in any other physical way, and
transmission or information storage and retrieval, electronic adaptation, computer software, or by similar
or dissimilar methodology now known or hereafter developed.
The use of general descriptive names, registered names, trademarks, service marks, etc. in this publication
does not imply, even in the absence of a specific statement, that such names are exempt from the relevant
protective laws and regulations and therefore free for general use.
The publisher, the authors and the editors are safe to assume that the advice and information in this book
are believed to be true and accurate at the date of publication. Neither the publisher nor the authors or
the editors give a warranty, expressed or implied, with respect to the material contained herein or for any
errors or omissions that may have been made. The publisher remains neutral with regard to jurisdictional
claims in published maps and institutional affiliations.

This Springer imprint is published by the registered company Springer Nature Switzerland AG
The registered company address is: Gewerbestrasse 11, 6330 Cham, Switzerland

Preface

Quantum mechanics is one of the fundamental theories of physics and has been tremendously successful at predicting and explaining the atomic and subatomic phenomena. Without quantum mechanics, there would be no modern world as we know it—most of the artifacts of what we consider to be a technologically advanced civilization only exist thanks to the fact that we managed to master quantum physics. Without it, there would be no electronic devices, no transistors, no lasers, no semiconductors, no MRI scanners, no GPS, no modern material science, no modern chemistry, and the list could go on for quite a while. John Preskill called the development of quantum theory "the crowning intellectual achievement of the last century".

Today, we are at the dawn of the quantum computing age, a multidisciplinary field that sits at the intersection of quantum physics, quantum information theory, computer science, and mathematics and may revolutionize the world of computing and software engineering.

In this book, we will introduce the basics of the theory of quantum computing, illustrated with runnable examples written in Q#, a quantum-specific programming language developed by Microsoft.

While quantum computing, compared to other areas of physics or computer science, is still a rather niche field, there is already a rich abundance of excellent books available on the market, approaching the subject matter from a wide array of interesting angles and written by some of the greatest minds in this field. Given this, one might ask a very valid question—what is the value proposition of this particular book then?

The short version of the answer would be, that in many aspects, this is the book I would have liked to read myself—though that is hardly a selling point. A longer answer to this question is not that straightforward, though hopefully one that will resonate with the reader. The major goal was to equip readers of various backgrounds, who are not yet familiar with quantum theory, with an approachable introductory quantum computing book that would offer a healthy mix of academic rigour, typically characterizing texts on topics related to quantum computing, with pragmatism that is so central to agile software development world. In particular, the hands-on, example-driven culture of learning new software engineering patterns or program-ming languages is often in strong juxtaposition to the strict abstract mathematical formalism one needs to go through when beginning to learn quantum computing. As such, in this book, all of the discussed theoretical aspects of quantum computing are always, as customary in software engineering publications, annotated with runnable code examples, providing the reader with two different angles—mathematical and programmatic—of looking at the same problem space.

It is my great hope that the book will be equally relevant to undergraduate students of physics and other related scientific disciplines, as well as to technically versed enthusiasts, such as software developers or self-learners, with background in other technologies.

Werner Heisenberg once wrote[1]

> The resolution of paradoxes of atomic physics can be accomplished only by further renunciation of old and cherished ideas.

From that perspective, quantum mechanics and, in the broader sense, quantum theory are quite possibly the most fascinating and intellectually gratifying areas of all science. The ability to understand and predict the world of the very small has been one of the greatest achievements of humanity and has not only given us an unprecedented technological growth but also propelled us on a quest to find out the true nature of reality more than any scientific theory before. In that sense, I sincerely hope that this book may also become an impulse sparking interest in quantum physics in a gentle and non-invasive fashion. The amazing attribute of quantum computing, after all, is that it provides anyone interested in the subject with an accessible, ready-made laboratory, where quantum phenomena can be experienced within a few keystrokes, using only a structured programming language. This, in turn, encourages experimen-tation and fosters our natural human curiosity and hopefully can translate into further exploration of the quantum domain, beyond quantum computing.

This book is structured into three major logical parts. The first one deals with foundational concepts, establishing the theoretical basis for the later parts of the study. It covers both the historical aspects of the quantum theory and quantum computing as well as offers a gentle introduction to quantum mechanics. This by no means is intended to be an exhaustive account, in fact quite the opposite, it specifically takes a strong focus on ideas relevant for quantum computing only, and solely using algebra. This contributes toward building up the foundational knowledge and helps establish a simple quantum mechanical context, without clouding the picture with complicated mathematical ideas—both of which, in turn, should help to smoothly navigate further chapters dedicated specifically to quantum computing. Finally, Part I also aims to familiarize the reader with the basic features of the Q# language and its related Quantum Development Kit.

---

[1] Werner Heisenberg, "The Physical Principles of the Quantum Theory", Scientific Review Papers, Talks, and Books Wissenschaftliche Übersichtsartikel, Vorträge und Bücher, Springer Berlin Heidelberg 1984.

Part II covers the core ideas of quantum computing, introducing the concepts of a qubit, quantum gates and quantum circuits. In addition to that, a series of important concepts from quantum information theory, in particular, entanglement, take center stage and are covered in-depth, followed by discussion about quantum key distribution and its various protocols.

Finally, Part III rounds the book off by covering the canon of most impactful quantum computational algorithms discovered thus far. This is of course a culmination point, one made possible by, and we can say that with a somewhat triumphal tone to it, taking advantage of everything covered in Parts I and II.

Such structure means that, in general, the book follows a certain progression path, intended to provide the most effective learning and reading experience. At the same time, the individual chapters, while building upon each other, are only loosely connected content-wise. Because of that, readers already familiar with quantum mechanics may skip Chap. 2, readers familiar with Q# and the QDK may omit Chap. 3 and readers familiar with basics of quantum computing can opt to jump past Chap. 4—the rest of the book will hopefully still provide value nonetheless.

All the code samples from this book are available on GitHub.[2] The quantum circuits have all been generated with the excellent quantikz[3] LaTeX package from Alastair Kay.

I hope that through the pages of this book, the reader will enjoy their first steps into the quantum computing world and will share the author's enthusiasm and admiration for this fascinating and incredibly diverse field of science.

Zürich, Switzerland   Filip Wojcieszyn
February 2022

---

[2] https://github.com/filipw/intro-to-qc-with-qsharp-book.
[3] https://ctan.org/pkg/quantikz.

Acknowledgements

First and foremost, I would like to thank my publisher Marina Forlizzi, who was so helpful, open-minded and did not hesitate to give me this amazing writing opportunity. Without her support for this project, this book would not have existed.

I am eternally grateful for the support I received from all my friends at the Microsoft Quantum team. Special thanks go to Matt Zanner who opened the doors for me and was always there when I needed something, to Cassandra Granade who has always been an inspiration and a role model to look up to and to Bettina Heim who has always been so helpful, approachable and patient with my inquiries and suggestions. I learnt a ton from all of you.

In addition to that, I cannot forget the great Q# community. In particular, I would like to say thanks to Sarah Kaiser who works tirelessly shaping it and has made it such a welcoming and caring place.

The University of the West of Scotland will always have a special place in my heart. Its School of Computing is the place that instilled in me the curiosity and inquisitiveness about the world and the laws governing it, especially from the computational angle. This has been one of my most cherished gifts, and a testament to how an academic institution can impact one's life.

There are also a few people I met on my professional path that I owe so much to, because they always unconditionally supported and believed in me—in particular Philipp Schneider in Switzerland and Chris Smillie in Canada. Thank you.

This book would not have happened without the fantastic support of my wife Maja and my daughter Lara, who put up with me buried in books and computer screens for a long, long, long time. I love you so much.

Contents

Part I

Chapter 1
Historical Background

We shall begin by covering the birth of quantum mechanics and the historical aspects of the development of quantum theory. In this initial chapter we will also contextualize the role of the famous EPR thought-experiment, the Einstein-Born intellectual debate that it triggered and the subsequent dramatic discovery of Bell's inequalities. A brief history of quantum computing follows, which will give us a first chance to familiarize ourselves with the key names and concepts in the field.

Overall, this chapter provides the historical frame of reference for the rest of the book, throughout which the topics and ideas mentioned in this introductory part are often revisited, refined and reflected upon—both algebraically and programmatically.

1.1 Historical Development of Quantum Theory

By the end of 19th century, most physicists considered the scientific description and understanding of the world to be complete. Light, with all of its wave properties was understood to be a form of electromagnetic radiation, while matter was known to be made of small indivisible units, called particles. Albert Michelson, a future Nobel Prize winner, best known for accurately measuring the speed of light, even proclaimed in a 1894 speech during the dedication of Ryerson Physical Laboratory that "it seems probable that most of the grand underlying principles have been firmly established" [50]. The next five decades showed that he could not have been more wrong.

In 1900, Max Planck, working on a blackbody radiation problem, introduced a postulate that energy is radiated in discrete packets, rather than in a continual spectrum. This radical idea, called quantization of energy, was initially a mere math-

© The Author(s), under exclusive license to Springer Nature Switzerland AG 2022
F. Wojcieszyn, *Introduction to Quantum Computing with Q# and QDK*, Quantum Science and Technology, https://doi.org/10.1007/978-3-030-99379-5_1

ematical trick that made the calculations finally work out.[1] In fact, Planck tried so hard to not depart from the classical ideas in physics, that his goal was to develop a theory in which the concept of quanta only appeared in the emission of energy, while absorption happened continually. His guess was right though, and for this discovery, Planck was nominated for Nobel Prize in physics in 1908, and ultimately went on to win it in 1918.

Albert Einstein adopted Planck's idea to explain the so-called photoelectric effect in 1905 [20]. It was the first one of the four papers that became known as "Annus Mirabilis"[2] papers, a creative rush of substantive scientific work that had never before been, and likely will never again be, contributed by a single individual in such a short period of time. Even more stunning is the fact that at that time Einstein was still not a full-time academic researcher but a patent clerk at the Swiss Patent Office in Bern. Einstein showed that metals emit electrons when light shines on them due to the interaction between the electrons that made up those metals and the *quanta of light* which ultimately became known as *photons*. This of course created another puzzle, and later became one of the key characteristics of the quantum theory—namely the wave-particle duality, as light suddenly could be reasoned about in two completely different, seemingly incompatible ways. For his work on the photoelectric effect Einstein received the Nobel Prize in physics in 1921.[3]

Not long after the successful quantization of light, Planck's quantum led to yet another scientific breakthrough—in 1913 Niels Bohr used the idea to provide the first successful description of the hydrogen atom. Bohr, using a mixture of classical and quantum physics, proposed that electrons can only occupy certain orbits around the nucleus, which he called stationary states, each one corresponding to a specific energy level. Electrons could only *jump* between these states by emitting or absorbing quantized energy. Despite the fact that this model managed to finally successfully describe the discrete frequencies that hydrogen atom emitted, the so-called spectral lines of the atom, the initial reception of Bohr's theory was reserved, if not cold. The main concern was that it was in violation of Maxwell's equations—under classical rules such model would simply not be stable. But there were physicists that immediately saw the brilliance behind the quantized atom and the need to depart from classical thinking. Ernst Rutherford, Bohr's mentor when his three part paper was published, was one of his most vocal supporters, while Albert Einstein was reported to have said "this is one of the greatest discoveries" [22] when he heard about Bohr's theory. Bohr's hydrogen atom description was a massive step forward in the understanding of atomic structure and ultimately led to him receiving a Nobel Prize in physics in 1922.[4]

---

[1] Helge Kragh tells an excellent background story of the Planck's discovery and calls him the "reluctant revolutionary" in [44].

[2] lat. miracle year.

[3] Remarkably, Einstein's crowning and most spectacular scientific achievement, the general theory of relativity, did not earn him a Nobel Prize.

[4] Bohr was nominated, among others, by Max Planck.

Another stunning discovery came in 1924, when Louis de Broglie showed in his doctoral thesis that it was not only light that exhibited wave-particle duality, but also matter. His doctoral examiners at the Sorbonne, in particular the external examiner Paul Langevin, were not convinced by his arguments and considered the conclusions of de Broglie to be too far fetching, so they sent the paper to Albert Einstein to obtain his second opinion. Einstein quickly recognized the significance of de Broglie's discovery and the brilliance in his line of thinking, which utilized Planck's quantum to link the particle's momentum to its wavelength and created a beautiful symmetry between light and matter. Einstein responded to Langevin that "he [de Broglie] has lifted a corner of the great veil" [1]. De Broglie's went on to win the 1929 Nobel Prize in physics for his work on the wave nature of electrons.

Despite these significant steps in developing the quantum theory by quantizing further areas of physics, there still remained fundamental problems—in particular, the orbital motion of electrons, a major flaw of Bohr's atom remained unanswered, as any attempt to describe the mechanics of particles failed.

The major breakthrough came in 1925 when an associate of Bohr at his institute in Copenhagen, Werner Heisenberg, published what effectively become the foundation of quantum mechanics. He started with classical equations of motion and mathematically reinterpreted them in a bold and counter-intuitive way. Heisenberg realized that in order to provide a predictive quantum theory he does not need a realist physical description of the quantum phenomena, and instead introduced a completely abstract mathematical formalism, which focused on "quantities that are in principle observable" [33]. In practice, it meant that we could speak of interactions between quantum objects and the measurement apparatus but not about the motion of those objects. It was a radical move, which led him to resolve the problem of the unstable electron orbits in Bohr's hydrogen atom—they could no longer be observed and thus there existed no classical ontology for them. To represent his quantum mechanics, Heisenberg used custom mathematical constructs of infinite-dimensional sets of complex variables, which was quickly reformulated into proper matrices[5] [14], resulting in this particular formulation of quantum mechanics becoming known as *matrix mechanics*.

On top of that, Heisenberg introduced *noncommutativity*[6] between observables, which was a completely novel idea in theoretical physics at the time. He realized that at quantum level, observable properties such as momentum and position should not commute, namely, we cannot measure both of them with arbitrary accuracy at the same time. This notion, which lies at the heart of quantum mechanics, became known as the *Heisenberg uncertainty principle*.

The departure from the idealized, Laplacian approach of classical physics was a profound, brilliant step. Arkady Plotnitsky reflected on these developments [55],

---

[5] Matrix theory was not particularly well known area of mathematics among physicists at the time. Heisenberg was not aware of the concept of matrices, but in what is undoubtedly a testament to his brilliance, guided by his intuition, he effectively ended up reinventing matrix mathematics as part of his work on quantum mechanics.

[6] Given two quantities P and Q, we can express noncommutativity as $P \times Q \neq Q \times P$.

by pointing out that Heisenberg, through his usage of matrix-like formalism, non-commutativity, and by bringing new type of variables into established theoretical physics, created foundations for bringing physics, mathematics and philosophy into a completely new relationship equilibrium, which "constitutes a truly extraordinary accomplishment, rivalling (...) those of Newton and Einstein".

Soon after Heisenberg's work was published, Paul Dirac restructured Heisenberg's mechanics further, into a much more mathematically coherent noncommutative algebra [19] and developed his unique braket notation which until today is the ubiquitous mathematical notation used in quantum physics.

Independently of Heisenberg, in 1926, Erwin Schrödinger formulated his own quantum theory based on a modified version of the classical wave equation, which became known as the Schrödinger equation, the foundation of the so-called *wave mechanics*. Contrary to Heisenberg, who embraced the fact that not all physical attributes of the quantum system, can, in principle, be observed at the same time, Schrödinger was guided by much different physical principles. He set off to restore classical, Newtonian motion description into quantum realm. His program was very quickly shown to be not a viable proposition as it could not work in a three dimensional classical reality—it required a high amount of dimensions, just like Heisenberg's matrix mechanics. Max Born, who worked together with Heisenberg on the formulation of quantum mechanics at the University of Göttingen, showed the same year [11] that Schrödinger's troubles of painting a classical picture with his wave equation stems from the fact that the function itself is not a real entity. Instead, he formulated the so-called *Born rule*, which can be used to convert an abstract wave function state into classical probability distributions. This allowed him to resolve the paradox of the highly dimensional aspect of Schrödinger's wave mechanics and reinterpret it back into a construct that fits into three-dimensional world. Born wrote [12]

> Either we use waves in spaces of more than three dimensions (...) or we remain in the three-dimensional space, but give up the single picture of the wave amplitude, and replace it by a purely mathematical concept (...) confined to the quantity $||\psi\rangle|^2$ (the square of the amplitude).

In the end it was John von Neumann who in 1932 showed beyond any doubt, that matrix and wave mechanics are mathematically equivalent as they are both "formulations of one and the same conceptual substratum" [42]. By that point the mathematical foundations were firmly established, and have not changed much since then—von Neumann's mathematical formalism is now the standard mathematical language of quantum mechanics.

The Born rule is now one of the postulates of quantum mechanics and has never been derived, nor violated empirically. It also provides a probabilistic framework for predicting quantum phenomena. According to Plotnitsky [55], from the philosophical standpoint, the Born rule "gave Schrödinger's mechanics the Heisenbergian epistemology", and effectively "sounded the death knell for Schrödinger's program of wave mechanics" in terms of the ontological character of the wave function itself that Schrödinger advocated for.

In 1927, in another breakthrough step in the development of quantum mechanics, Wolfgang Pauli formalized the mathematics behind non-relativistic theory of spin, and discovered the Pauli matrices as representations of the spin operators. Pauli, just like Heisenberg was at one point Born's assistant at the University of Göttingen, and just like Heisenberg went on to become a giant of quantum theory. His contributions to modern quantum theory were immense,[7] spearheaded by the formulation of the Pauli exclusion principle [53], published when he was only 25 years old.[8] In 1928, Pauli was appointed a Professor of Theoretical Physics at ETH Zürich, where, after a brief time in the USA during the 2nd World War, he went on to build the physics institute, have an illustrious scientific career and become a towering figure in the history of this prestigious university that has produced 21 Nobel prize winners—one of them being Albert Einstein himself.[9]

In 1939, Pauli's assistant at ETH, Markus Fierz, in his habilitation degree [26], formulated the spin-statistics theorem, which was later refined further by Pauli himself [52]. Spin, which is a property of each elementary particle, represents one of two types of angular momentums that can be found in quantum mechanics—intrinsic angular momentum. The value of the spin is a discrete number and, in general, all particles can be divided into two families—those with half-integer spin, called *fermions*, and those with full integer spin, called *bosons*. Their role in nature is fundamentally different, with fermions making up what we consider ordinary matter (quarks or electrons), and bosons (photons, gluons or for example Higgs boson) acting as force carriers. Regardless whether it is half- or full-integer spin, it can take one of only two possible in a specific spatial dimension. From that perspective particle spin makes up the simplest possible quantum system, and is very useful in describing the basics of quantum information theory, as well as introducing quantum computers. We shall return to those concepts in Chap. 2.

Werner Heisenberg received the Nobel Prize in physics in 1932 for "the discovery of quantum mechanics", upon nomination from both Albert Einstein and Niels Bohr, while Dirac and Schrödinger shared the 1933 prize for the further advancements of the quantum theory, or as it was formulated in the award "for the discovery of new productive forms of atomic theory." Pauli received the Nobel prize in physics in 1945 for the Pauli exclusion principle—also after the nomination from Albert Einstein. Max Born received Nobel prize in 1954 for "for his statistical interpretation of the wave function".

While the abstract mathematical formalism proved to provide excellent reflection of experimental data and be an incredibly accurate predictor of new phenomena,

---

[7] Pauli was famously difficult to work with, but his scientific contributions are spectacular. Max Born once remarked that Pauli "was a genius, comparable only to Einstein himself" and that as a scientist he was "perhaps even greater than Einstein" [13].

[8] Since quantum mechanics was dominated by young scientists, mostly in their 20s—Heisenberg, Pauli, Dirac, de Broglie and others—it quickly became known in the scientific circles as *Knabenphysik*, German for "boy physics".

[9] It is therefore of no surprise that today, the ETH Institute for Theoretical Physics, which was established upon Pauli's arrival at the university, is located at the Wolfgang-Pauli-Strasse 27 in Zürich.

the counter-intuitive probabilistic nature, bizarre rules, confusing epistemology and complete lack of ontology of quantum mechanics have troubled some of the greatest physicists of the 20th century. Most famously, quantum mechanics has deeply upset Albert Einstein, who could not accept the probabilistic core of quantum mechanics related to the Born rule and Heisenberg's uncertainty principle, and was convinced the ultimate nature of reality is fundamentally deterministic. Einstein, in a 1926 letter to Max Born [13], remarked

> The theory produces a good deal but hardly brings us closer to the secret of the Old One. I am at all events convinced that He does not play dice.

Apparently, as recalled by Werner Heisenberg [35], during a 1927 Solvay conference Niels Bohr replied to Einstein

> (...) it cannot be for us to tell God, how He is to run the world.

1.2   Entanglement and the Debate About True Nature of Reality

The probabilistic nature of quantum mechanics was in stark contrast to classical physics, which, since Newtonian ideas about the "clockwork universe", was considered to be fundamentally deterministic. This led to considerable difficulties in trying to capture the nature of reality within the quantum theory. The response from the founding fathers of quantum mechanics was a philosophical interpretative framework known as the *Copenhagen interpretation*, championed primarily by Niels Bohr[10] and several other physicists that were close to him, such as Heisenberg, Pauli or Born. In line with the Born rule, in the view adopted by them, the wave function of a quantum object does not describe the nature or reality in any direct way—it merely exists as a mathematical tool. Moreover, we are not even allowed to reason about the existence of that quantum object until it's actually measured. In other words, it is impossible disassociate the reality responsible for the quantum phenomena from the measurement itself, as we can only observe trace effects of quantum objects on the measurement apparatus and only infer their existence that way. This is fundamentally different from classical macro-scale physics, where systems, their state and behavior can be objectively described and observed.

As one might imagine, ontological problems started to pile up quickly. Erwin Schrödinger proposed [59] a thought experiment that highlighted a paradox when considering an interaction between a quantum object and a macroscopic object. He imagined a cat, locked in a steel chamber which was linked to a poison. Additionally, a radioactive substance was placed next to the chamber, with a Geiger counter. As soon as one of the radioactive atoms decayed, the Geiger counter would trigger

---

[10] It needs to be emphasized that there never existed one single coherent "Copenhagen interpretation", not even from Bohr himself, whose point of view continuously evolved over time. However, we can speak about a general "spirit of Copenhagen" of related ideas.

the mechanism that released the poison and killed the cat. Schrödinger pointed out that the fate of the cat is tied to the atomic process of radioactive decay, described by quantum theory and that according to the mathematical formalism of quantum mechanism, and the interpretative framework of Copenhagen, the cat is both dead and alive at the same time, until it is observed.

Schrödinger's cat has since gone to establish itself as one of the symbols of quantum theory and has gained a firm position in popular culture. The main difficulty it highlighted, called the *measurement problem*, has remained unsolved (at least in a universally accepted way) to this day, giving rise to a wide range of competing interpretational frameworks of quantum mechanics, rivaling the original philosophical groundwork from Bohr.[11]

Quantum mechanics is the only fundamental theory of nature in which probabilities are postulated "ab initio" [9], a state of affairs that many physicists objected to and opposed Bohr's point of view. Most famously, Albert Einstein claimed that while quantum theory advanced our understanding of atomic-level phenomena, its "statistical nature will turn out to be transitory" [57].

In 1935 Einstein, along with Boris Podolsky and Nathan Rosen [21] described an apparent problem in the predictions provided by the mathematics of quantum mechanics. Through a thought experiment around the so-called "EPR pair"—two particles that interact with each other—they realized that quantum mechanics allows those particles to enter an *entangled state*.[12] They pointed out that in such entangled state, through measuring a certain physical property of the first member of the pair, such as position x_1, the measurement result of the same physical property of the second particle, x_2, can be predicted with certainty. Repeating this process again, now measuring a different property such as momentum p_1, we can predict with probability equal to one the value of p_2. Those became know as the so-called EPR correlations. What follows, according to the EPR argument, is that we can simultaneously ascribe two incompatible conjugate properties to a quantum object, thus violating the uncertainty principle around which the new quantum theory was built. In this way, they argued, quantum mechanics was an incomplete as a description of physical reality.

Einstein followed the EPR paper with a famous lengthy exchange with Bohr, who provided an initial response to the paper. In the resulting intellectual scuffle, Bohr pointed out that the EPR reasoning is generally incorrect due to the irreducible role of the measuring instruments in quantum theory. According to Bohr, the logic followed by the EPR paper required two separate experimental arrangements—each one only allowing to predict one set of property values for a *different* particle pair— and is impossible to realize within one experiment for a *single* particle pair. He then objected to inference needed between those two separate experimental setups. In

---

[11] These topics, while formidably interesting, are not relevant to the content of this book. A curious reader is referred to many fine publications covering the so-called *quantum foundations* in depth, the conversations of Schlosshauer with eminent physicists [58], the recent work of Laloë [46] and Plotnitsky [55, 56] being excellent places to begin with.

[12] The EPR paper, while conceptually describing the phenomenon of entanglement, did not use that term—it was introduced later in the same year by Erwin Schrödinger, in his "cat-paradox" paper.

other words—we never learn the exact conjugate property values of a *specific* EPR pair, which, according to Bohr, was rendering the argument as fundamentally invalid.

As a consequence, Einstein shifted the focus of the argument to the apparent non-locality of quantum mechanics—violation of special relativity theory, since the EPR correlations appeared to be some sort of force or communication between the two particles that was instant, and thus, seemingly, faster than the speed of light. Locality was the fundamental property of reality that Einstein was not prepared to give up. Any possible violation of local realism was unacceptable to him, and he famously proceeded to calling the EPR correlations *spooky action at a distance* [21].

Since violation of determinism and local realism were out of question, his preferred alternative to quantum mechanics was a local realist theory based on the ontological concept of *local hidden variables*. In such scheme, there would exist real properties of quantum objects which the mathematical formalism of quantum mechanics does not take into account, and which, in Einstein's view, pre-determined the quantum correlations between the particles making up an EPR pair, and to which the perceived probabilistic nature of quantum theory would be attributable. This would have not only restored locality into quantum theory but also reintroduced the Newtonian determinism into the quantum processes—after all, what appeared random to us, would have been predetermined by those hidden variables.

A simplified variant of the EPR thought experiment, highlighting the non-local aspects of quantum mechanics was proposed by David Bohm in 1951 [10].[13] Bohm defined a source emitting an entangled positron-electron pair, which could be represented e.g. by a decaying neutral pion. Each particle making up the emitted pair flies off towards separate observers—Alice and Bob. When Alice measures the spin of her particle along the specific spatial direction, she will randomly get a result of $\frac{1}{2}$ or $-\frac{1}{2}$, but as soon as that happens, due to the fact that the decayed pion had spin zero and that conservation of angular moment applies, it is immediately known that Bob will be guaranteed to produce opposite measurement result upon his measurement along the same direction. Additionally, quantum mechanics predicts that the entangled particles could become spatially separated by enormous distances, even light years, before getting measured by the observers—and the correlations would still apply.

Einstein never found a locality-obeying hidden variables model that could serve as an alternative to quantum mechanics, and the program ultimately failed. It was ruled out completely by John S. Bell, who, using Bohm's ideas, disproved the "local hidden variables" approach to quantum mechanics using a brilliant method of comparing classical correlations and correlations between entangled quantum objects [5] in what became known as Bell's inequality, or more broadly, the Bell's theorem. Because of the reliance on classical correlations, this inequality will always be satisfied if one takes the local realist stance suggested by Einstein. On the other hand, Bell's inequality will get violated for certain quantum mechanical entangled states. In other words, Bell managed to set a mathematical upper bound on the predictions made by hidden variable theories, and then showed that there are quantum mechan-

---

[13] Bohm's variant has been widely used, refined further and restated in various forms in quantum literature over the years, including a very digestible and approachable explanation by Griffiths [30].

ical phenomena whose occurrence cannot be explained using a local realist hidden variable approach.

A few years after Bell's work was published, Clauser, Horne, Shimony, and Holt introduced the so-called CHSH inequality [16], which was a generalization of Bell's ideas based on treating the correlation functions as averages over hidden variables.

It took almost two decades to experimentally confirm Bell's theorem, which was first done by a group led by Alain Aspect at the École supérieure d'optique in Orsay [3]. Many other experimental variants of verifying Bell's inequality followed, closing subsequent loopholes that were still possible in the earlier experimental configurations [32, 36]. At this stage we are confident to claim that entangled particles exhibit correlations that are beyond explanation of any local hidden variables theory.

Bell's theorem is one of the most puzzling aspects of the nature of reality—Henry Stapp even pronounced it to be *the most profound discovery in science* [62]. While there is no consensus among physicists and philosophers about what are the ontological and epistemological consequences of Bell's theorem, by proving Einstein's local realist position wrong, it has nevertheless profoundly impacted the field of quantum foundations, forcing physicists to effectively choose to give up either the notion of locality or abandon local realism [30].

It is clear that quantum entanglement occupies space right at the very heart of the confusing landscape of quantum theory. The problem is amplified by the usage of the term *entanglement* itself—the quantum behavior is named after a known, relatable, visualizable and inherently local classical concept. Unfortunately, quantum entanglement with its non-local correlations is nothing like the classical, macro-scale term, leading to a sudden and violent breakdown of our intuition. These deceptive linguistic traps are characteristic for quantum theory and have often been studied in quantum theoretical literature over the years. MacKinnon [47] refers to this as "using language beyond the limits of normal usage."

This makes any attempts of building a philosophical model of reality based on quantum theory a rather grim activity. On one hand, the sophisticated mathematical scheme used to perform quantum calculations does not lend itself well to be used as description or representation of the underlying reality. On the other, the natural language used to theorize about quantum phenomena is troubled by invalid or counter-intuitive analogies and burdened by language constructs that are tainted with classical connotations. The result of that is that the cognitive barriers around quantum mechanics are high, if not insurmountable. The entire situation is somewhat unfortunate, but also, for the most part, inescapable. It is also not entirely new, as it was already apparent to the founders of quantum mechanics. Heisenberg noted [34]

> (...) it is very difficult to modify our language so that it will be able to describe atomic processes, for words can only describe things of which we can form mental pictures.

The phenomenon of entanglement contributes to the massive divide between quantum and classical theoretical thinking, up to a point where, as it was the case with Einstein, it drifts so far away from our acceptable rational boundaries that it becomes difficult, or even impossible to accept it as being the true nature of reality. We will explore the algebra behind entanglement and Bell's theorem further in Chaps. 2 and 5, and rely on entanglement in almost every quantum computing algorithm covered in this book.

1.3   History of Quantum Computing

Computer science owes a great deal of debt to Alan Turing, who in the 1930s, through his work on topics such as problem universality and general computability, laid out the mathematical foundations for the entire field of computation theory. Most famously, he defined a generalized mathematical model of computation that later became known as a *Turing machine* [63].

Turing described this imaginary machine as a computing device capable of replicating mathematical states and being operated by a human. Even more critically, Turning defined the so called *computable numbers*, which would be calculable on the universal machine. A Turing machine can be thought of as a first general example of a central processing unit—a critical theoretical milestone that ultimately led to the creation of the first computers.

Around the same time, independently of Turing, an American mathematician Alonzo Church was working on the similar problem space—effective computability problems and, more generally, on the mathematical field of recursion. In fact, it was actually Church who coined the term *Turing machine* after reading Turing's paper. Together, they contributed towards establishing a concept of the nature of problems (functions) whose output is effectively calculable, which became known as the *Church–Turing thesis*:

> Every function which would naturally be regarded as computable can be computed by the universal Turing machine.

Historically, the theory of quantum computing dates back to early 1980s, when Paul Benioff theorized about building a classical Turing machine that was a microscopic quantum mechanical model [6]. Roughly at the same time, in the Soviet Union, completely independently of Benioff and unbeknownst to the western scientific world, the brilliant Russian physicist Yuri Manin proposed a concept of building a quantum computer [48].

Soon after, in 1981, Richard Feynman delivered a famous lecture [24] at the 1st MIT Physics and Computation conference, in which he argued that in order to be able to simulate Nature accurately, we should not use classical computing, but a completely new computation model based on quantum mechanics. He said

> I'm not happy with all the analyses that go with just classical theory, because Nature isn't classical, dammit, and if you want to make a simulation of Nature, you'd better make it quantum mechanical.

This lecture is widely regarded as the moment that jump started the area of quantum computing. Feynman further refined these ideas further in [25].

As it turned out, quantum computing found itself at odds with the Church–Turing thesis very quickly. The thesis is a conjecture—it is not directly refutable, but for half a century it underpinned our understanding of what computational models are and has been treated as an axiom. However, in 1985, David Deutsch published a groundbreaking paper [18] in which he redefined the Church–Turing thesis into the following stronger principle:

> Every finitely realizable physical system can be perfectly simulated by a universal model computing machine operating by finite means.

Deutsch's reasoning was that the term *naturally*, used by Turing, created mathematical ambiguity, while his reformulation referred to objective concepts of *physical system*, and was therefore compatible with measurement theory. He then proceeded to argue that his reformulated principle carried a statement much stronger than the original version—in fact, it was so strongly formulated that at this point not even the Turing machine was compatible with it. Deutsch pointed out that perfect simulations of continuous problems, for example those in classical dynamics, are not possible on a Turing machine, because those effectively rely on a finite ways of preparing input. In addition, along the lines of thinking of Richard Feynman, classical computational models rely on states which are perfectly known and fully described at any computation stage—which in principle is incompatible with quantum mechanics. What follows, naturally, is that accurate and reliable simulation of quantum systems is de facto impossible on classical computers.

Deutsch addressed all of these concerns by defining, for the first time, what he called a *universal quantum computer* and described its computational model—using *qubits*,[14] quantum bits, instead of bits as the indivisible unit of information. It was much different from classical computing and because it was based on the laws of quantum mechanics, the only one satisfying his strong formulation of the Church–Turing principle [18]. We shall return to that paper in Chap. 7, as it also contained the first problem statement and an algorithm in which quantum computer required less logical steps than classical device to reach the solution.

In 1982, Nick Herbert [37] suggested a paradox based on copying unknown quantum states. He imagined Alice and Bob, being physically separated from each other, holding large amount of EPR pairs in their possessions. By cloning his part of the EPR pair a large amount of times, Bob could abuse the apparent non-local effects of quantum correlations—obtaining the information in an instant, faster than it would take for light to travel between Alice and Bob [8]. This scheme could then be used by them to exchange information at superluminal velocity, without being bound by classical restrictions. In other words, if cloning of unknown quantum states was permitted, one could, in principle, use entanglement for superluminal or backwards in time communication.

---

[14] The term was not used by Deutsch, it was introduced later by Schumacher [60] in his study on interpreting quantum states as information. It took a while for it to fully catch on and become universally accepted, with some physicists preferring a simpler version *qbit*. With his usual light-heartedness, David Mermin even referred to the usage of qubit over qbit as "preposterous" [49].

Many physicists intuitively knew this was wrong [54], because it would of course be in conflict with the special theory of relativity, as it could lead us to valid arrangements where effect would precede cause. And indeed, it was in response to Herbert's paper that the so called *no-cloning theorem* was conceived. In 1982, William Wootters and Wojciech Zurek derived the theorem in a letter sent to Nature [66].[15] They used linearity of quantum mechanics as a mechanism to prove that replicating quantum states in a generic way is forbidden. We will go through this proof in Chap. 2.

In [43] Richard Jozsa expanded the original proof with a stronger version of no-cloning theorem by analyzing what supplementary information is needed in order to successfully clone a quantum state. He then concluded that the state must be supplemented with so much information, that the clone would be possible to be generated from the additional information alone, independently of the initial state. In other words, the original state offers no contribution to cloning.

From a larger perspective no-cloning theorem carries certain ontological weight as well. The question about the ability to copy an arbitrary quantum state becomes rather a question about the true nature of quantum state, and whether we can even speak about its existence in a classical sense prior to the act of measurement—a question, which Wootters and Zurek answered with a "no" [67].

The no-cloning theorem not only precluded the possibility of faster-than-light communication, but it also helped kick start a fascinating area of research, namely quantum cryptography and quantum key distribution. Naturally, if unknown quantum states cannot be cloned, we can already speak about a foundational basis for a secure communication protocol. And indeed, quantum cryptography as a separate field was born in 1984, when Charles Bennett and Gilles Brassard used no-cloning theorem as the basis for their work on the theoretical model for the first quantum key distribution protocol [7]. We will discuss it in detail in Chap. 6.

In the first part of the 1990s, two of the most important algorithms for quantum computers were developed. First, in a bombshell publication that shook not only the physics community, but also the computer science world, Peter Shor of IBM showed that quantum computers can offer exponential speed up when factoring large integers [61]. The idea that factoring large numbers is a computationally unsolvable task underpins most of the modern cryptography, and the possibility that quantum computers might be able to solve that problem efficiently was a dramatic development, that attracted a lot of attention to the field. In 1996, Lov Grover of Bell Labs developed a quantum search algorithm through an unsorted data set that provided a quadratic speedup over the classical counterparts [31]. We are going to dedicate a lot of room to both of these algorithms in Chap. 7.

Despite very promising early results on the algorithmic front, during its first two decades quantum computing remained mostly a theoretical field, as the hardware development proved to be very challenging. The first physical realizations of quantum computers were presented at the end of 1990s, independently of each other, by two separate research groups [17, 27], using a technology called *nuclear magnetic*

---

[15] The title "A single quantum cannot be cloned", was contributed by John Wheeler [54].

resonance, with a computational volume of two qubits. This type of quantum computer architecture was also used for a first physical realization of Grover's search algorithm [15], and to the first implementation of Shor's factorization algorithm, when number 15 was successfully factored [64]. The experimental progress somewhat stagnated afterwards, particularly due to the fact the nuclear magnetic resonance turned out to be rather difficult to scale. It was not until the last several years, that the hardware development pace in quantum computing industry picked up rapidly again, and there are a few reasons for that.

First of all, there are now multiple different physical realizations of quantum hardware, most promising of which are arguably superconducting tunnel junctions and trapped ions [28]. In 2021 a group around the University of Science and Technology of China demonstrated their impressive 62-qubit programmable quantum computer called "Zu Chongzhi" based on a superconducting processor [29]. The same machine, in an improved configuration was later used to achieve *quantum supremacy*[16]—solve a task which is computationally impossible to solve on even the fastest supercomputer currently available [68]. Another interesting novel approach to building quantum computing devices are silicon qubits, which may end up benefitting from the existing semiconductor microchip solutions [65].

These technologies attract a lot of funding from both state actors [45] and venture capital [23], which in turn further fuels innovation and entrepreneurship spirit. As a result, quantum hardware is no longer a unique domain of academia, and through the involvement and funding from both public and private sectors the devices are getting more powerful, stable and accessible.

Secondly, the rise of cloud computing changed the perspective on what it means to operate quantum hardware. It is now possible to get access to quantum hardware via the cloud, which dramatically lowers the barriers of entry for companies and individuals interested in building quantum software. It also incentives hardware manufacturing, which, by becoming part of cloud platforms, can be serviced more easily and benefit from much larger demand resulting in tighter feedback loop between quantum software authors, quantum researchers and hardware producers. A landmark moment happened in 2016, when IBM launched their, partly freely available, "IBM Q Experience" cloud platform for quantum computing [38]. Initially equipped with a single 5-qubit machine, it contains, at the time of writing, 20+ machines and has played an important role in the growth and popularization of the quantum computing field. IBM has been very aggressive in the their quantum investments, and unveiled IBM Quantum System One, the world's first integrated quantum computing system, in 2019, to underpin their cloud offering. As of early 2022, the largest quantum machine available in the IBM cloud was their 65-qubit machine codenamed "Hummingbird", while a 127-qubit quantum processor unit codenamed "Eagle" was unveiled at the end of 2021 [41] and will soon become a feature of the cloud platform too. IBM's quantum computing roadmap [40] aims for a spectacular increase

---

[16] The original quantum supremacy barrier was broken by Google's Sycamore processor in 2019 [2], though the task computed by Zu Chongzhi is about 6 orders of magnitude more difficult. There have also been questions raised around the complexity of the task Sycamore solved [39].

in the quantum volume of the IBM Q platform's quantum hardware capabilities each year, with a massive 1121-qubit quantum processor units, codenamed "Condor", to become available as part of IBM Quantum System Two in 2023.

Other software giants are not staying put too. Amazon and Microsoft are both offering their own cloud computing quantum services on their cloud platforms, AWS and Azure. As of 2022, Microsoft Azure Quantum integrates two third-party trapped ion hardware startups Quantinuum and IonQ, and simultaneously heavily invests in research [51] and their own first-party hardware development.

Finally, there has been a lot of promising developments in the quantum software space, bridging the gap between theorists, hardware engineers and software engineers. Historically quantum software development required direct manipulation of the lowest level primitives responsible for controlling and transforming the quantum state of the system. On top of that, it was necessary for algorithm authors to actively engage in quantum error correction activities, which was hardly productive. This is, of course, in stark contrast to classical software engineering, where typically many levels of abstraction—firmwares, SDKs, libraries, operating systems and such—exist between the raw CPU instructions and the software developers building a program, stabilizing the overall experience and providing a very convenient, high-level environment for building software. This is now changing, as quantum computing is maturing; several convenient quantum-specific programming languages and frameworks have come to existence in recent years. Additionally, mature software concepts such as the emergence of the idea of quantum firmware [4] help tremendously as well, because even though the fidelity of quantum hardware is improving, there are still decoherence problems and computation errors that need to be dealt with, and having a robust abstraction layer sitting between the software and the hardware makes the life of algorithm and program authors considerably easier.

In fact, the theme of this book will be to look at the quantum theory through the lenses of the Q# programming language, a high level programming language specifically tailored for quantum programming, and its corresponding Quantum Development Kit from Microsoft. We will go through their features in Chap. 3 and rely on them for all code examples in this book. IBM has also developed a very vibrant open source community around Qiskit,[17] its Python framework for quantum computing, while Google AI Quantum Team started Cirq,[18] another Python framework for quantum development.

References

1. Abragam, A. (1988). Louis Victor Pierre Raymond De Broglie, 15 August 1892 - 19 March 1987. *Biographical Memoirs of Fellows of the Royal Society, 34,* 21–41.
2. Arute, F., Arya, K., Babbush, R., Bacon, D., Bardin, J., Barends, R., et al. (2019). Quantum supremacy using a programmable superconducting processor. *Nature, 574,* 505–510.

---

[17] https://qiskit.org.

[18] https://github.com/quantumlib/Cirq.

3. Aspect, A., Grangier, P., & Roger, G. (1982). Experimental realization of Einstein-Podolsky-Rosen-Bohm Gedankenexperiment: A new violation of Bell's inequalities. *Physical Review Letters, 49*(Jul), 91–94.
4. Ball, H., Biercuk, M. J., & Hush, M. R. (2021). Quantum firmware and the quantum computing stack. *Physics Today, 74*(3), 28–34.
5. Bell, J. S. (1964). On the Einstein Podolsky Rosen paradox. *Physics Physique Fizika, 1*(Nov), 195–200.
6. Benioff, P. (1980). The computer as a physical system: A microscopic quantum mechanical Hamiltonian model of computers as represented by Turing machines. *Journal of Statistical Physics, 22*(05), 563–591.
7. Bennett, C. H., & Brassard, G. (2014). Quantum cryptography: Public key distribution and coin tossing. *Theoretical Computer Science, 560*(Dec), 7–11.
8. Bera, R. K. (2020). *The amazing world of quantum computing.* Singapore: Springer.
9. Blanchard, P., & Fröhlich, J. (Eds.). (2015). *The message of quantum science. Attempts towards a synthesis.* Berlin: Springer.
10. Bohm, D. (1989). *Quantum theory (Dover books on physics).* New York: Dover Publications.
11. Born, M. (1926). Zur Quantenmechanik der Stossvorgänge. *Zeitschrift für Physik, 37*(12), 863–867.
12. Born, M. (1935). *Atomic physics.* London: Blackie and Son.
13. Born, M. (2004). *Born-Einstein letters, 1916-1955: Friendship, politics and physics in uncertain times.* London: Palgrave Macmillan.
14. Born, M., & Jordan, P. (1925). Zur Quantenmechanik. *Zeitschrift für Physik, 34*(1), 858–888.
15. Chuang, I., Gershenfeld, N., & Kubinec, M. (1998). Experimental implementation of fast quantum searching. *Physical Review Letters, 80*, 3408–3411.
16. Clauser, J. F., Horne, M. A., Shimony, A., & Holt, R. A. (1969). Proposed experiment to test local hidden-variable theories. *Physical Review Letters, 23*(Oct), 880–884.
17. Cory, D. G., Fahmy, A. F., & Havel, T. F. (1997). Ensemble quantum computing by NMR spectroscopy. *Proceedings of the National Academy of Sciences, 94*(5), 1634–1639.
18. Deutsch, D. (1985). Quantum theory, the Church-Turing principle and the universal quantum computer. *Proceedings of the Royal Society of London. Series A, Mathematical and Physical Sciences, 400*(1818), 97–117.
19. Dirac, P. A. M., & Fowler, R. H. (1925). The fundamental equations of quantum mechanics. *Proceedings of the Royal Society of London. Series A, Containing Papers of a Mathematical and Physical Character, 109*(752), 642–653.
20. Einstein, A. (1905). über einen die Erzeugung und Verwandlung des Lichtes betreffenden heuristischen Gesichtspunkt. *Annalen der Physik, 322*(6), 132–148.
21. Einstein, A., Podolsky, B., & Rosen, N. (1935). Can quantum-mechanical description of physical reality be considered complete? *Physical Review, 47*(May), 777–780.
22. Eve, A. S. (1939). *Rutherford: Being the life and letters of the Rt. Hon. Lord Rutherford, O.M.* London: Macmillan.
23. Feder, T. (2020). Quantum computing ramps up in private sector. *Physics Today, 73*(11), 22–25.
24. Feynman, R. P. (1982). Simulating physics with computers. *International Journal of Theoretical Physics, 21*(6), 467–488.
25. Feynman, R. P. (1985). Quantum mechanical computers. *Optics News, 11*(2), 11–20.
26. Fierz, M. (1939). Über die relativistische Theorie kräftefreier Teilchen mit beliebigem Spin. *Helvetica Physica Acta, 12*(Jan), 3–37.
27. Gershenfeld, N. A., & Chuang, I. L. (1997). Bulk spin-resonance quantum computation. *Science, 275*(5298), 350–356.
28. Gibney, E. (2020). Quantum computer race intensifies as alternative technology gains steam. *Nature, 587*(7834), 342–343.
29. Gong, M., Wang, S., Zha, C., Chen, M.-C., Huang, H.-L., Wu, Y., et al. (2021). Quantum walks on a programmable two-dimensional 62-qubit superconducting processor. *Science, 372*(6545), 948–952.

30. Griffiths, D. J. (2018). *Introduction to quantum mechanics* (3rd ed.). Cambridge: Cambridge University Press.
31. Grover, L. K. (1996). A fast quantum mechanical algorithm for database search. In *Annual ACM Symposium on Theory of Computing* (pp. 212–219). ACM.
32. Handsteiner, J., Friedman, A. S., Rauch, D., Gallicchio, J., Liu, B., Hosp, H., et al. (2017). Cosmic Bell test: Measurement settings from milky way stars. *Physical Review Letters, 118*(6).
33. Heisenberg, W. (1925). über quantentheoretische Umdeutung kinematischer und mechanischer Beziehungen. *Zeitschrift Für Physik, 33*(12), 879–893.
34. Heisenberg, W. (1984). *The physical principles of the quantum theory* (pp. 117–166). Berlin: Springer.
35. Heisenberg, W. (1989). *Encounters with Einstein: And other essays on people, places, and particles*. Princeton: Princeton University Press.
36. Hensen, B., Bernien, H., Dréau, A. E., Reiserer, A., Kalb, N., Blok, M. S., et al. (2015). Loophole-free Bell inequality violation using electron spins separated by 1.3 kilometres. *Nature, 526*(7575), 682–686.
37. Herbert, N. (1982). FLASH—A superluminal communicator based upon a new kind of quantum measurement. *Foundations of Physics, 12*(12), 1171–1179.
38. IBM. (2016). IBM makes quantum computing available on IBM cloud to accelerate innovation. https://www-03.ibm.com/press/us/en/pressrelease/49661.wss. Retrieved 13 May 2021.
39. IBM. (2019). On "Quantum supremacy". https://www.ibm.com/blogs/research/2019/10/on-quantum-supremacy/. Retrieved 12 Feb 2022.
40. IBM. (2020). IBM's roadmap for building an open quantum software ecosystem. https://www.ibm.com/blogs/research/2021/02/quantum-development-roadmap/. Retrieved 13 May 2021.
41. IBM. (2021). IBM unveils breakthrough 127-qubit quantum processor. https://newsroom.ibm.com/2021-11-16-IBM-Unveils-Breakthrough-127-Qubit-Quantum-Processor. Retrieved 28 Dec 2021.
42. Jammer, M., & of Physics, American Institute. (1989). *The conceptual development of quantum mechanics*. History of modern physics and astronomy series. Los Angeles: Tomash Publishers.
43. Jozsa, R. (2002). A stronger no-cloning theorem.
44. Kragh, H. (2000). Max Planck: The reluctant revolutionary. *Physics World, 13*(12), 31–36.
45. Kramer, D. (2021). Israel has become a powerhouse in quantum technologies. *Physics Today, 74*(12), 24–27.
46. Laloe, F. (2019). *Do we really understand quantum mechanics?* Cambridge: Cambridge University Press.
47. MacKinnon, E. (2012). *Interpreting physics. Language and the classical/quantum divide.* Netherlands: Springer.
48. Manin, Y. (1980). Computable and uncomputable. *Sovetskoye Radio.*
49. Mermin, N. D. (2007). *Quantum computer science. An introduction.* Cambridge: Cambridge University Press.
50. of Chicago, University. (1896). *Annual register 1896.* Chicago: University of Chicago.
51. Pauka, S. J., Das, K., Kalra, R., Moini, A., Yang, Y., Trainer, M., et al. (2021). A cryogenic CMOS chip for generating control signals for multiple qubits. *Nature Electronics, 4*(January), 64–70.
52. Pauli, W. (1940). The connection between spin and statistics. *Physical Review, 58*(Oct), 716–722.
53. Pauli, W. (2010). On the connexion between the completion of electron groups in an atom with the complex structure of spectra.
54. Peres, A. (2003). How the no-cloning theorem got its name. *Fortschritte der Physik, 51*(45), 458–461.
55. Plotnitsky, A. (2010). *Epistemology and probability: Bohr, Heisenberg, Schrödinger, and the nature of quantum-theoretical thinking.* New York: Springer.
56. Plotnitsky, A. (2016). *The principles of quantum theory, from Planck's Quanta to the Higgs Boson.* New York: Springer International Publishing.

57. Purrington, R. D. (2018). *The heroic age: The creation of quantum mechanics, 1925–1940.* Oxford: Oxford University Press.
58. Schlosshauer, M. (Ed.). (2011). *Elegance and Enigma.* The Frontiers collection. Berlin: Springer.
59. Schrödinger, E. (1935). Die gegenwärtige Situation in der Quantenmechanik. *Naturwissenschaften, 23*(48), 807–812.
60. Schumacher, B. (1995). Quantum coding. *Physical Review A, 51*(Apr), 2738–2747.
61. Shor, P. W. (1997). Polynomial-time algorithms for prime factorization and discrete logarithms on a quantum computer. *SIAM Journal on Computing, 26*(5), 1484–1509.
62. Stapp, H. P. (1975). Bell's theorem and world process. *Il Nuovo Cimento B (1971-1996), 29*(2), 270–276.
63. Turing, A. M. (1937). On computable numbers, with an application to the Entscheidungsproblem. *Proceedings of the London Mathematical Society, s2-42*(1), 230–265.
64. Vandersypen, L. M. K., Steffen, M., Breyta, G., Yannoni, C. S., Sherwood, M. H., & Chuang, I. L. (2001). Experimental realization of Shor's quantum factoring algorithm using nuclear magnetic resonance. *Nature, 414*(6866), 883–887.
65. Warren, A., & Economou, S. E. (2022). Silicon qubits move a step closer to achieving error correction. *Nature, 601*, 320–322.
66. Wootters, W. K., & Zurek, W. H. (1982). A single quantum cannot be cloned. *Nature, 299*(5886), 802–803.
67. Wootters, W. K., & Zurek, W. H. (2009). The no-cloning theorem. *Physics Today, 62*(2), 76–77.
68. Zhu, Q., Cao, S., Chen, F., Chen, M.-C., Chen, X., Chung, T.-H., et al. (2021). Quantum computational advantage via 60-qubit 24-cycle random circuit sampling.

Chapter 2
Basics of Quantum Mechanics

Throughout this book, we shall be focusing on quantum mechanics exclusively from the perspective of quantum computation theory, thus specifically avoiding a whole range of challenging topics, such as the Schrödinger equation or the Hamiltonian operator. We will instead operate entirely on—one would hope!—more familiar algebraic concepts of vector-based states and matrices, and approach them solely from the angle of their usefulness with regards to programming quantum computers.

Over the next few dozen pages we will cover the basics of quantum mechanics, with the formidable goal of equipping ourselves with theoretical foundations needed before proceeding into the field of quantum computing from Chap. 4 onwards.

2.1 Quantum State

Hilbert space is a complex vector space, which, as a mathematical concept, was developed in the early 20th century by David Hilbert. Hilbert space can accommodate any number of dimensions, making it an invaluable tool in many branches of mathematics and physics. The usage of Hilbert spaces in quantum mechanics was introduced by John von Neumann and is the formalism that we still use today. We denote the vector space as $\mathbb{C}^n$, where n indicates the number of dimensions. For example $\mathbb{C}^2$ indicates that all elements of this vector space are two dimensional.

In quantum mechanics, the preferred notation when dealing with vectors is the Dirac notation, introduced by Paul Dirac [8], also known as the *braket notation*. A ket is simply a column vector written with a certain label—for example $|\psi\rangle$ is a vector denoting a state labeled ψ, without indicating anything about the dimensions of this vector. We are generally free to assign any labels to the kets, for example $|photon\rangle$ or $|electron\rangle$ are perfectly valid names, though greek letters are most prevalent for

© The Author(s), under exclusive license to Springer Nature Switzerland AG 2022
F. Wojcieszyn, *Introduction to Quantum Computing with Q# and QDK*, Quantum Science and Technology, https://doi.org/10.1007/978-3-030-99379-5_2

unknown states. In quantum computing, as we will see in Chap. 4, the labels $|0\rangle$ and $|1\rangle$, due to correspondence with classical bit values, are most commonly used.

The simplest possible quantum state $|\psi\rangle$ is described by a two-dimensional vector in the complex Hilbert space $\mathbb{C}^2$ and can be written as

$$|\psi\rangle = \begin{bmatrix} a \\ b \end{bmatrix} \tag{2.1}$$

where a and b are complex numbers. Complex numbers are made up of a real an imaginary components using the format

$$a + xi \tag{2.2}$$

where a and x are real numbers, and $i = \sqrt{-1}$. The set of complex numbers is a superset of real numbers—every real number is a complex number with the imaginary component equal to 0.

A bra, on the other hand, is the conjugate transpose of the ket and in the braket notation looks like a mirror of a ket.

$$\langle\psi| = |\psi\rangle^\dagger = \begin{bmatrix} a^\star & b^\star \end{bmatrix} \tag{2.3}$$

In linear algebra, conjugate transpose is denoted by the $\dagger$ symbol and takes the column vector to a row vector, while additionally changing the sign of the imaginary component of the complex numbers. For example:

$$\begin{bmatrix} a + xi \\ b + yi \end{bmatrix}^\dagger = \begin{bmatrix} a - xi & b - yi \end{bmatrix} \tag{2.4}$$

Vectors in Hilbert space are equipped with the inner product, which allows length and vector angles to be defined. The inner product (also known as the scalar product) in the Dirac notation is called a braket, and can be expressed as follows:

$$\langle\varphi|\psi\rangle = \begin{bmatrix} c^\star & d^\star \end{bmatrix} \begin{bmatrix} a \\ b \end{bmatrix} = c^\star a + d^\star b \tag{2.5}$$

In order for a vector to be a valid state vector in quantum mechanics, it needs to be normalized, which means it must be a unit vector—have a length of 1. The term comes from the fact that in linear algebra we refer to the length of the vector as *norm*. The norm is the square root of the inner product.

$$\|a\| = \sqrt{\langle a|a\rangle} \tag{2.6}$$

Since the length is always a real positive number, this means that the inner product of the quantum state vector with its conjugate transpose must also be equal to 1.

$$\langle \psi | \psi \rangle = 1 \tag{2.7}$$

Any vector could be converted into a unit vector if it is divided by its norm. At that point it becomes a valid description of a quantum state.

$$|\tilde{\psi}\rangle = \frac{|a\rangle}{\|a\|} = \frac{|a\rangle}{\sqrt{\langle a | a \rangle}} \tag{2.8}$$

2.2   Two-Level Quantum System

Now that we know how the express a quantum state, let us see how we can use it to describe the state of a simple two-level quantum system. By a two-level quantum system we understand a system that has two distinct states, which we can be denoted using arbitrary ket labels. For example, when measuring an electron spin along the z-axis, we could speak of spin direction up or down, and therefore use the states $|\uparrow\rangle$ and $|\downarrow\rangle$. Similarly, a photon going through a polarizer might end up in one of two possible polarization states such as $|\nwarrow\rangle$ or $|\nearrow\rangle$.

In order to have a tangible real world reference, we will assume working with spins for the time being, but all the mathematics here is equally applicable to any other two-level quantum systems. Electrons have spin value which, when measured along the z-axis, is equal to $+\frac{\hbar}{2}$ for spin up and $-\frac{\hbar}{2}$ for spin down, where $\hbar$ is the reduced Planck constant.

It means that spin values are non-continuous, but discrete—spin can only assume one of two possible values, which also gives it a very useful designation of being the simplest possible quantum system. Spin can be measured in any spatial direction, such as along the z-axis, the x-axis or the y-axis.

Since the two states of the two-level quantum system must always be distinguishable from each other, in linear algebra the vectors corresponding to them must be linearly independent (orthogonal), otherwise the vectors might overlap with each other in some situations. Orthogonality of two vectors can be expressed as their inner product being equal to zero.

$$\langle \uparrow | \downarrow \rangle = 0 \tag{2.9}$$

Since we already established that quantum states are represented by unit vectors, so vectors whose norm is equal to 1, these two orthogonal vectors can be called orthonormal, and together they make up a so called *orthonormal basis*. We indicate that the vector pair forms a basis by writing them together in curly braces: $\{|\uparrow\rangle, |\downarrow\rangle\}$. We shall refer to it as the *Z-basis*.

There are infinitely many possible orthogonal unit vector pairs and we are free to choose any such pair to represent two distinct quantum states. However there is a pair of them that are particularly useful because they make the related mathematical calculations easier.[1] Those are the vectors for which a and b are equal to 0 and 1, and are shown in Eq. 2.10. We shall use them to represent spin up and spin down along the z-axis.

$$|\uparrow\rangle = \begin{bmatrix} 1 \\ 0 \end{bmatrix} \qquad |\downarrow\rangle = \begin{bmatrix} 0 \\ 1 \end{bmatrix} \tag{2.10}$$

2.2.1 Superposition

The basis vectors can be used to decompose any vector in two-dimensional space into a linear combination of them. Therefore, in our case, where the basis we chose is denoted by $\{|\uparrow\rangle, |\downarrow\rangle\}$, any general quantum state $|\psi\rangle$ can be written as:

$$|\psi\rangle = \begin{bmatrix} a \\ b \end{bmatrix} = a \begin{bmatrix} 1 \\ 0 \end{bmatrix} + b \begin{bmatrix} 0 \\ 1 \end{bmatrix} = a |\uparrow\rangle + b |\downarrow\rangle \tag{2.11}$$

It is easy to see from Eq. 2.11, that when $a = 1$, then $b = 0$, $|\psi\rangle$ is in the state $|\uparrow\rangle$. Conversely, when $a = 0$, then $b = 1$ and the state $|\uparrow\rangle$ is aligned with $|\downarrow\rangle$.

$$|\psi\rangle = \begin{bmatrix} 1 \\ 0 \end{bmatrix} = 1 \begin{bmatrix} 1 \\ 0 \end{bmatrix} + 0 \begin{bmatrix} 0 \\ 1 \end{bmatrix} = |\uparrow\rangle$$

$$|\psi\rangle = \begin{bmatrix} 0 \\ 1 \end{bmatrix} = 0 \begin{bmatrix} 1 \\ 0 \end{bmatrix} + 1 \begin{bmatrix} 0 \\ 1 \end{bmatrix} = |\downarrow\rangle \tag{2.12}$$

Since the vector length is restricted to be equal to 1, the values of a and b are restricted by the normalization condition:

$$|a|^2 + |b|^2 = 1 \tag{2.13}$$

Notice that if we restricted ourselves to real numbers only, so used the vector space $\mathbb{R}^2$ instead of $\mathbb{C}^2$, then the condition could be written without taking the absolute values prior to squaring them.

$$a^2 + b^2 = 1 \tag{2.14}$$

However, in quantum mechanics, the vectors are complex and for complex numbers, it is possible to still have a negative number after squaring it, because of the imaginary number i.

$$i^2 = -1 \tag{2.15}$$

---

[1] There is another reason too, that will become apparent very soon once we start exploring the concept of observables.

In all the cases when a and b are non-zero, we have a state that is called a *superposition* of the two basis states. While a and b are permitted to assume any values as long the normalization condition from Eq. 2.13 is fulfilled, when $a = b$ we speak about *uniform superposition*. Such example is shown in Eq. 2.16.

$$|\psi\rangle = \begin{bmatrix} \frac{1}{\sqrt{2}} \\ \frac{1}{\sqrt{2}} \end{bmatrix} = \frac{1}{\sqrt{2}} \begin{bmatrix} 1 \\ 0 \end{bmatrix} + \frac{1}{\sqrt{2}} \begin{bmatrix} 0 \\ 1 \end{bmatrix} = \frac{1}{\sqrt{2}} \left(|\uparrow\rangle + |\downarrow\rangle \right) \tag{2.16}$$

So far we were only working with the $\{|\uparrow\rangle, |\downarrow\rangle\}$ basis. However, as we have already mentioned, we have full freedom in choosing the bases we would like to use when reasoning about a given quantum state. For example, the following two vectors form a perfectly valid orthonormal basis $\{|\leftarrow\rangle, |\rightarrow\rangle\}$ and logically would correspond to spin measurements along the x-axis in the three dimensional space. Hence, we can refer to it as the *X-basis*.

$$|\leftarrow\rangle = \frac{1}{\sqrt{2}} \begin{bmatrix} 1 \\ 1 \end{bmatrix} \qquad |\rightarrow\rangle = \frac{1}{\sqrt{2}} \begin{bmatrix} 1 \\ -1 \end{bmatrix} \tag{2.17}$$

Let us now consider the superposition state $|\psi\rangle$ from Eq. 2.16. As it turns out, that state, a superposition in $\{|\uparrow\rangle, |\downarrow\rangle\}$ basis is actually equal to $|\leftarrow\rangle$ when expressing this state as decomposed in the basis $\{|\leftarrow\rangle, |\rightarrow\rangle\}$

$$|\psi\rangle = \frac{1}{\sqrt{2}} \left(|\uparrow\rangle + |\downarrow\rangle \right) = \begin{bmatrix} \frac{1}{\sqrt{2}} \\ \frac{1}{\sqrt{2}} \end{bmatrix} = |\leftarrow\rangle \tag{2.18}$$

Conversely, both of the basis states $|\uparrow\rangle$ and $|\downarrow\rangle$ are in a uniform superposition with respect to $\{|\leftarrow\rangle, |\rightarrow\rangle\}$ basis. For example

$$|\uparrow\rangle = \begin{bmatrix} 1 \\ 0 \end{bmatrix} = \frac{1}{\sqrt{2}} \begin{bmatrix} \frac{1}{\sqrt{2}} \\ \frac{1}{\sqrt{2}} \end{bmatrix} + \frac{1}{\sqrt{2}} \begin{bmatrix} \frac{1}{\sqrt{2}} \\ -\frac{1}{\sqrt{2}} \end{bmatrix} = \frac{1}{\sqrt{2}} \left(|\leftarrow\rangle + |\rightarrow\rangle \right) \tag{2.19}$$

This tells us something very profound. Superposition is *basis dependent*. Such conclusion should not be surprising though, as it is a direct consequence of the vector-based mathematical formalism of quantum mechanics. The same vector state can be represented as linear combination of infinitely many other vector pairs forming other orthonormal bases.

In other words, a state is always in superposition with respect to certain bases and not in superposition to some other basis. If a system is in a definite basis state with respect to one basis, and simultaneously in a uniform superposition with respect to another basis, we would call such two bases *mutually unbiased bases*.

2.2.2 Bloch Sphere

We already mentioned that the quantum mechanical mathematical formalism relies on complex numbers. Let us revisit that briefly by examining the spin states $|\uparrow\rangle$ and $|\leftarrow\rangle$ again. The spin state $|\uparrow\rangle$ is a $+\frac{\hbar}{2}$ *vertical* spin, which implies direction along the three dimensional z-axis. On the other hand, the spin state $|\leftarrow\rangle$ is a $+\frac{\hbar}{2}$ *horizontal* one, along the x-axis.

Looking at their vector representations, since they are only described with real numbers and can be plotted on a two-dimensional plane, there appears to be a $45° = \frac{\pi}{4}$ angle between them, which is seemingly incompatible with the fact that in three-dimensional space we have a $90° = \frac{\pi}{2}$ angle between z- and x-axes.

This tells us that the rotation by an angle θ in real space, needs to be equivalent to a rotation by $\frac{\theta}{2}$ angle in the quantum state space. We can use that[2] to rewrite the general state of a two-level quantum system, for now in the x-z plane, using an arbitrary value of the θ angle, instead of the usual coefficients a and b

$$|\psi_{xz}\rangle = \cos\left(\frac{\theta}{2}\right)|\uparrow\rangle + \sin\left(\frac{\theta}{2}\right)|\downarrow\rangle \tag{2.20}$$

If we now substitute θ with $0° = 0\pi$, $90° = \frac{\pi}{2}$ and $180° = \pi$, we obtain

$$|\psi(0)\rangle = \cos(0)|\uparrow\rangle + \sin(0)|\downarrow\rangle = |\uparrow\rangle$$

$$\left|\psi\left(\frac{\pi}{2}\right)\right\rangle = \cos\left(\frac{\pi}{4}\right)|\uparrow\rangle + \sin\left(\frac{\pi}{4}\right)|\downarrow\rangle = \frac{1}{\sqrt{2}}\left(|\uparrow\rangle + |\downarrow\rangle\right)$$

$$|\psi(\pi)\rangle = \cos\left(\frac{\pi}{2}\right)|\uparrow\rangle + \sin\left(\frac{\pi}{2}\right)|\downarrow\rangle = |\downarrow\rangle \tag{2.21}$$

which is exactly what we were expecting.

This still only covers two of the spatial dimensions though, but it is also possible to have the spin measured along the y-axis. Consequently, we need to introduce an extra angle ϕ to cover the missing coordinate plane. We will need a new orthonormal basis for the y-axis, which will be made up of the orthogonal "in" and "out" states $\{|\circlearrowleft\rangle, |\circlearrowright\rangle\}$ and this is where we need to finally introduce complex numbers. We will refer to it as the *Y-basis*.

$$|\circlearrowleft\rangle = \frac{1}{\sqrt{2}}\begin{bmatrix} 1 \\ i \end{bmatrix} \qquad |\circlearrowright\rangle = \frac{1}{\sqrt{2}}\begin{bmatrix} 1 \\ -i \end{bmatrix} \tag{2.22}$$

Similarly to Eq. 2.20, we can express a two-level quantum system for any arbitrary value of the ϕ angle with regards to the x-y plane as

$$|\psi_{xy}\rangle = \frac{1}{\sqrt{2}}|\uparrow\rangle + \frac{e^{i\phi}}{\sqrt{2}}|\downarrow\rangle \tag{2.23}$$

---

[2] The reasoning in this section is inspired by the excellent presentation of Kok [13].

The reason why we use the exponent e instead of the angle directly would be that quantum state vectors must be normalized to a fixed length of 1, and this is the way to guarantee this.

Geometrically speaking, and taking into account Eqs. 2.20 and 2.23 to produce

$$|\psi\rangle = \cos\left(\frac{\theta}{2}\right)|\uparrow\rangle + e^{i\phi}\sin\left(\frac{\theta}{2}\right)|\downarrow\rangle \qquad (2.24)$$

we arrive at the conclusion that because of the complex number formalism, any two level quantum state can be plotted on a three-dimensional sphere with a fixed radius of 1, where every surface point represents one of the infinitely many unique quantum states for the system. Orthogonal states are then defined by two points on the surface of the sphere that are exactly opposite from each other. This sphere is called the Bloch sphere, and only works for two-level quantum states—for larger systems, as we will learn later, we reach more dimensions and our visualization capabilities break down. We will use the methodology introduced here later on when discussing Bell's theorem in Sect. 2.9.

The final thing we need to do is that we need to set the possible limits for values of both θ and ϕ. They cannot both range the entire $360° = 2\pi$, as that would cover the entire three-dimensional sphere twice, so one of them must be limited to $180° = \pi$. While this is completely arbitrary, the conventional approach in quantum mechanics is to do it for θ, giving us the allowed ranges (in radians)

$$0 \leq \theta \leq \pi \qquad 0 \leq \phi < 2\pi \qquad (2.25)$$

Such mathematical formalism and the Bloch sphere visualization model leads us to one final observation. All the bases for two-level quantum states are simply made up of two antipodal points on the surface of the Bloch sphere. We can always change from one basis to another by performing a rotation around the Bloch sphere—for example, the X-basis is actually the Z-basis rotated around the y-axis by $\pm\frac{\pi}{4}$. This knowledge will come in very useful once we start working with quantum computations.

For a long time the necessity for using complex numbers in quantum mechanics has not been formally proven, instead it has merely been postulated and has simply always provided accurate predictions of the experimental outcomes. Nevertheless, it has been suggested that it might be possible to give standard quantum theory mathematical description using real numbers only. However, in 2022, two Chinese research groups [5, 15] performed ingenious experiments that refuted the conjecture that complex numbers are not required in the standard formulation of quantum mechanics [16].

2.3 Born Rule

While a two-level quantum state can be described by a continuous range of values for both a and b complex numbers, we can only observe one of the two basis states of our chosen orthonormal basis. This concept lies at the heart of quantum mechanics—in order to learn something about the quantum system, we need to perform a *measurement*, and contrary to classical physics, observations do influence the quantum state itself.

If the quantum state aligns with one of the two basis states, then the measurement result is deterministic. However, if the state is in the superposition with respect to the measurement basis, then the measurement is probabilistic—and the rules for this probabilistic result are governed by the so-called *Born rule*. The rule, named after Max Born who introduced a probabilistic interpretation of the wave function [2], is an axiom in quantum mechanics; it has never been derived yet it has always managed to perfectly align with experimental results. The Born rule can be concisely expressed in linear algebra.

First, the inner product between the unknown state $|\psi\rangle$ and the basis state $|\uparrow\rangle$ is:

$$\langle\uparrow\,|\psi\rangle = a\langle\uparrow\,|\,\uparrow\rangle + b\langle\uparrow\,|\,\downarrow\rangle = a \cdot 1 + b \cdot 0 = a \tag{2.26}$$

The Born rule in this case tells us that the classical probability of finding $|\psi\rangle$ in the state $|\uparrow\rangle$ is the absolute square of the inner product between the current quantum system state and the tested measurement state, which also is one of the basis states:

$$Pr(\uparrow) = |\langle\uparrow\,|\psi\rangle|^2 = |a|^2 \tag{2.27}$$

Based on this, we can calculate that the probability of finding the measured state in the basis state $|\downarrow\rangle$ is:

$$\langle\downarrow\,|\psi\rangle = a\langle\uparrow\,|\,\uparrow\rangle + b\langle\uparrow\,|\,\downarrow\rangle = a \cdot 0 + b \cdot 1 = b$$
$$Pr(\downarrow) = |\langle\downarrow\,|\psi\rangle|^2 = |b|^2 \tag{2.28}$$

This also explains the normalization condition from Eq. 2.13. Because $|a|^2$ and $|b|^2$ represent classical probabilities of two mutually exclusive possible measurement outcomes, their sum must be equal to one. Due to the fact that a and b could so easily be converted into classical probabilities, we refer to them as *probability amplitudes*.

The Born rule can be experimentally confirmed using a wide range of quantum experiments, one of the most famous of which is the Stern-Gerlach experiment. In the experiment, the electrons are prepared with a spin along the z- or x-axis, fired from a source, travel through a magnetic field created by two magnets and hit a fluorescent screen, which becomes a de-facto act of measurement, subject to the Born rule. Spin is a form of an angular momentum, so the magnetic field deflects the electrons traveling towards the screen in one of two possible directions away from the original path along which they were fired. It is the positioning of magnets that determines whether the measurement is performed along the z-axis (electrons are deflected up or down) or along the y-axis (electrons are deflected left or right). Algebraically this corresponds to a Z-basis $\{|\uparrow\rangle, |\downarrow\rangle\}$ or X-basis $\{|\leftarrow\rangle, |\rightarrow\rangle\}$ measurement.

If the initial spin is prepared along the z-axis, for example in the state $|\uparrow\rangle$, and the magnets are also set up for the z-axis measurement, then all of the electrons will only be deflected up. This is consistent with the description initially covered in Eq. 2.27:

$$Pr(\uparrow) = |\langle \uparrow \mid \uparrow \rangle|^2 = 1$$
$$Pr(\downarrow) = |\langle \downarrow \mid \uparrow \rangle|^2 = 0 \qquad (2.29)$$

Conversely, if the initial spin is prepared along the x-axis, for example in the state $|\leftarrow\rangle$, and the magnets are still set up for the z-axis measurement, we have, according to the Born rule maximum uncertainty for receiving a result $|\uparrow\rangle$ or $|\downarrow\rangle$—after all, we are measuring along the z-axis, so those are the only two measurement outcomes possible. We can describe this algebraically by expressing the state $|\leftarrow\rangle$ as the linear combination of both $|\uparrow\rangle$ or $|\downarrow\rangle$

$$|\leftarrow\rangle = \frac{1}{\sqrt{2}} \left(|\uparrow\rangle + |\downarrow\rangle \right) \qquad (2.30)$$

which happens to be the exact same state as introduced in Eq. 2.16. Since both a and b are equal to $\frac{1}{\sqrt{2}}$, the probabilities according to the Born rule are

$$Pr(\uparrow) = \left| \frac{1}{\sqrt{2}} \right|^2 = \frac{1}{2}$$
$$Pr(\downarrow) = \left| \frac{1}{\sqrt{2}} \right|^2 = \frac{1}{2} \qquad (2.31)$$

This leads us to conclude that a state in a *uniform superposition* with respect to the Z-basis $\{|\uparrow\rangle, |\downarrow\rangle\}$ is equally likely to produce any of the two orthogonal Z-basis states upon measurement. Consequently, in the Stern-Gerlach apparatus, this will manifest itself by having half of the electrons deflected up, and half of them down.

Similar effect can be achieved by rotating the magnets in the apparatus to deflect the electrons left or right only, thus performing an X-basis $\{|\leftarrow\rangle, |\rightarrow\rangle\}$ measurement, while preparing the initial electron spin along the z-axis in a state $|\uparrow\rangle$ or $|\downarrow\rangle$.

We already referred to the Z-basis and X-basis as mutually unbiased basis pair. We are now in a position to more formally state that if we measure a system, prepared in a basis state of one of the mutually unbiased bases, using *the other* of the two mutually unbiased bases, the observation will yield any possible measure outcome with equal probability. More colloquially one can refer to such measurement result as a perfect coin toss underwritten by the laws of Nature.

2.4 State Evolution

2.4.1 Unitary Transformations

In quantum mechanics, arbitrary transformations of the quantum state are not possible. Instead, the time evolution of a quantum system happens through a series of *unitary transformations*. Quite remarkably, unitarity is the only requirement for state evolution in quantum mechanics. Any unitary matrix $\mathbf{U}$ is a valid quantum state transformation, and is referred to as an operator.

$$\mathbf{U}\,|\psi_1\rangle = |\psi_2\rangle \tag{2.32}$$

A linear transformation $\mathbf{U}$ is said to be unitary when it produces an identity matrix $\mathbf{I}$

$$\mathbf{I} = \begin{bmatrix} 1 & 0 \\ 0 & 1 \end{bmatrix} \tag{2.33}$$

when multiplied by its Hermitian adjoint—a generalized version of conjugate transpose $\mathbf{U}^*$ from linear algebra—$\mathbf{U}^\dagger$.

$$\mathbf{U}^\dagger\mathbf{U} = \mathbf{U}\mathbf{U}^\dagger = \mathbf{I} \tag{2.34}$$

We could also say that differently—a matrix is unitary when its complex conjugate is also its own inverse.

$$\mathbf{U}^\dagger = \mathbf{U}^{-1} \tag{2.35}$$

This also means that every operation is by definition reversible, since applying the adjoint after the transformation will return the vector back to its original state.

$$\mathbf{U}\,|\psi_1\rangle = |\psi_2\rangle$$
$$\mathbf{U}^\dagger\,|\psi_2\rangle = |\psi_1\rangle \tag{2.36}$$

A matrix can also be equal to its own complex conjugate. It is then its own unitary inverses such that

$$\mathbf{U} = \mathbf{U}^\dagger \tag{2.37}$$

Such matrices are called Hermitian or self-adjoint operators, and applying the same operator twice in a row would recover the original quantum state.

$$\mathbf{U}\,|\psi_1\rangle = |\psi_2\rangle$$
$$\mathbf{U}\,|\psi_2\rangle = |\psi_1\rangle \tag{2.38}$$

After applying unitary transformations, the transformed state remains normalized, since unitary transformations do preserve the inner products between vectors. From the geometrical perspective, we can picture such transformations as conserving parallelity and only performing vector rotations. In quantum mechanics, as we already stated earlier, the length of the state vector must be equal to one.

$$\langle \psi | \psi \rangle = \langle \psi | \mathbf{U}^\dagger \mathbf{U} | \psi \rangle = 1 \tag{2.39}$$

Naturally, if we apply the same transformation to two different state vectors, the inner product between them is also preserved. This is very important because it guarantees that two orthogonal vectors remain orthogonal after the unitary transformation. For example, if the vectors $|u\rangle$ and $|v\rangle$ are orthogonal:

$$\langle u | v \rangle = 0 \tag{2.40}$$

then they remain orthogonal after applying $\mathbf{U}$:

$$\langle u | \mathbf{U}^\dagger \mathbf{U} | v \rangle = 0 \tag{2.41}$$

Overall, quantum state evolution is completely deterministic as we are able to calculate the system state at any point. Because each unitary transformation is reversible, no information is ever lost.

2.4.2   Beam Splitter

Cleve et al. [7] proposed using the basic algebra of a beam splitter and its interference effects as a gentle way to introduce quantum computational concepts. We shall take a similar approach here and introduce such simplified way of modelling a light beam travelling through a beam splitter to illustrate quantum state evolution. This theory will help us a lot when discussing quantum gates in Chap. 4.

Let us consider an optical experiment where light can travel from the light source along a specifically angled path towards a beam splitter. In the first variant, light will arrive from the top-left of the beam splitter, as shown in Fig. 2.1.

The beam splitter will allow 50% of light to go through and continue along the line it was taking, while the other 50% will be reflected at a 90° degree from its original path, so that the beam approaching from top-left would be heading in a top-right direction. Since light is an electromagnetic wave, there is nothing particularly surprising about it being split into two light waves heading off in two separate directions.

In the second variant, the light source will emit light approaching from the bottom-left of the beam splitter, as depicted in Fig. 2.2. In this case the beam of light is coming from bottom-left of the beam splitter, which will still allow half of the light to pass through, while the other half will be reflected towards the bottom-right direction, again at a 90° degree from its original path.

Now, let us imagine that the intensity of light is turned down so much that instead of a constant beam, the light source is emitting individual photons, one by one. In order to model these photons algebraically, we will need to define the orthonormal basis which will allow us to describe their quantum states as they approach the beam splitter. Because we are free to choose our ket labels however we want, we will

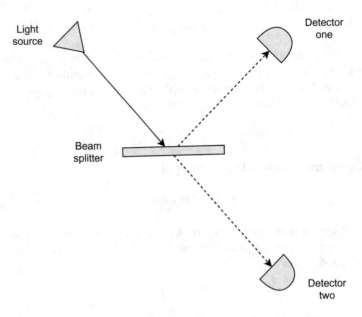

Fig. 2.1 Beam splitter with light approaching from the top-left

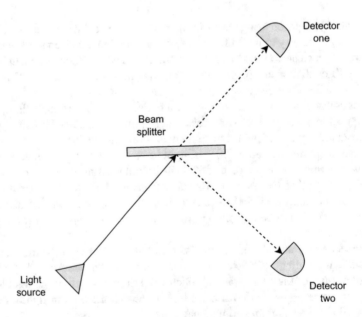

Fig. 2.2 Beam splitter with light approaching from the bottom-left

denote the state of the photon approaching along the top-left beam path as $|u\rangle$, while
the state of the photon approaching the beam splitter from the bottom-left path will
be represented by $|v\rangle$. Since we are still dealing with a two-level quantum state, the

system can only be measured to be in one of these two states. Because of that, we can ascribe these labels to the detectors as well, so in our setup detector one will correspond to measuring $|u\rangle$, while detector two to measuring $|v\rangle$.

The basis states written in a standard vector format are still going to be equal to the Z-basis vectors from Eq. 2.10.

$$|u\rangle = \begin{bmatrix} 1 \\ 0 \end{bmatrix} \quad |v\rangle = \begin{bmatrix} 0 \\ 1 \end{bmatrix} \tag{2.42}$$

We can model a simplified [4] balanced beam splitter using a unitary operator $\mathbf{U_B}$ shown in Eq. 2.43. It would take an input basis state $|u\rangle$ or $|v\rangle$ and create a uniform superposition of them, where one of the output beams, depending on the approach path, would be phase shifted by π.

$$\mathbf{U_B} = \frac{1}{\sqrt{2}} \begin{bmatrix} 1 & 1 \\ 1 & e^{i\pi} \end{bmatrix} \tag{2.43}$$

We could obviously calculate that

$$e^{i\pi} = -1 \tag{2.44}$$

which leads us to the so called *Hadamard matrix*

$$\mathbf{U_B} = \frac{1}{\sqrt{2}} \begin{bmatrix} 1 & 1 \\ 1 & -1 \end{bmatrix} \tag{2.45}$$

As a result, when the photon approaches from the top-left, we can express the process of going through the beam splitter as

$$\mathbf{U_B}|u\rangle = \frac{1}{\sqrt{2}} \begin{bmatrix} 1 & 1 \\ 1 & -1 \end{bmatrix} \begin{bmatrix} 1 \\ 0 \end{bmatrix} = \frac{1}{\sqrt{2}} \begin{bmatrix} 1 \\ 1 \end{bmatrix} = \frac{1}{\sqrt{2}} \left(|u\rangle + |v\rangle \right) \tag{2.46}$$

Conversely, the photon approaching along the bottom-left path can be expressed as

$$\mathbf{U_B}|v\rangle = \frac{1}{\sqrt{2}} \begin{bmatrix} 1 & 1 \\ 1 & -1 \end{bmatrix} \begin{bmatrix} 0 \\ 1 \end{bmatrix} = \frac{1}{\sqrt{2}} \begin{bmatrix} 1 \\ -1 \end{bmatrix} = \frac{1}{\sqrt{2}} \left(|u\rangle - |v\rangle \right) \tag{2.47}$$

We can easily see that the shift of the phase by π manifested itself with a change in sign of the superposition terms—from $(-)$ to $(+)$. This however does not impact the classical probabilities. Upon observation, such as placing detectors along both possible paths, as shown in both of Figs. 2.1 and 2.2, the system state will collapse to a measurement readout of either $|u\rangle$ or $|v\rangle$, according to the Born probability rule—in this case with 50% probability for both reflected and not reflected cases—and we will physically discover the photon at one of the detectors only. As a result, in both of these experimental setups the detectors flash randomly.

The remarkable thing about such approach to quantum state evolution is that in this formalism we could say that the beam splitter $\mathbf{U_B}$ *both reflected and not reflected* the photon at the same time—since it created a superposition of both reflected and pass-through states.

2.4.3 Mach-Zehnder Interferometer

So far both of the outgoing beam paths leaving the beam splitter were pointing in different directions, but we can force them to cross again by adding two mirrors, which would reflect them towards each other. We can then add another beam splitter at the location where they would cross. Such set up is called the Mach-Zehnder interferometer, and is shown in Fig. 2.3. If we use the superposition state from Eq. 2.46 as the input state into the second beam splitter, we will arrive at

$$\mathbf{U_B}\left(\frac{1}{\sqrt{2}}(|u\rangle + |v\rangle)\right) = \begin{bmatrix} 1 \\ 0 \end{bmatrix} = |u\rangle \tag{2.48}$$

This tells us that the application of the second beam splitter reverses the transformation done by the first one. In general, this should not be surprising—in quantum mechanics all unitary transformations are reversible after all. What is interesting here is that it is the same transformation, applied twice in a row, ended up undoing itself—the reason for such behavior is that the operator $\mathbf{U_B}$ used to describe the beam splitter is not only unitary, but also Hermitian—it is its own inverse, as we defined in Eq. 2.37.

If we now place detectors along the two paths leaving the second beam splitter, as done in Fig. 2.3, there will be no probabilistic outcome anymore. Because the quantum state is now a definite $|u\rangle$ state, with probability amplitude equal to one,

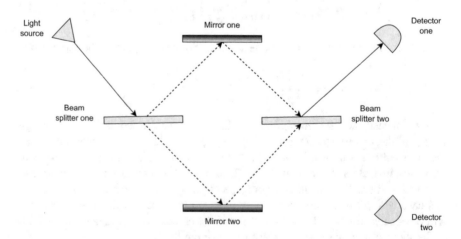

Fig. 2.3 Mach-Zehnder interferometer with light approaching from the top-left

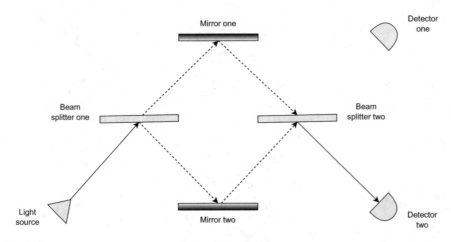

Fig. 2.4 Mach-Zehnder interferometer with light approaching from the bottom-left

only one of the final detectors will ever light up. In other words, since we ascribe states $|u\rangle$ or $|v\rangle$ to the detectors as well, based on the initial state of the system, which is determined by the initial angle of the light approaching the first beam splitter, we can deterministically say which specific detector would fire all the time. In this case, only detector one flashes.

The situation would be analogous if the input state into the second beam splitter is the one from Eq. 2.47, namely the initial beam of light entering the Mach-Zehnder interferometer approached from the bottom-left in state $|v\rangle$, as shown in Fig. 2.4

The corresponding quantum state can then be described as

$$\mathbf{U_B}\left(\frac{1}{\sqrt{2}}(|u\rangle - |v\rangle)\right) = \begin{bmatrix} 0 \\ 1 \end{bmatrix} = |v\rangle \tag{2.49}$$

where the final state is $|v\rangle$ and only detector two, corresponding to $|v\rangle$, will fire.

Yet again, this effect is not particularly surprising when the light is emitted as a continuous beam, because the observed effect, called *interference*, in this case destructive interference caused by the difference in relative phase, is an integral aspect of wave mechanics and can be observed in any waves, including those visible on the water surface. However, at the quantum level, the same phenomenon still arises when the light source emits individual photons. This of course is a highly counter-intuitive situation, because the way we can interpret it, is that the photon, even though it is indivisible, after the first beam splitter ends up taking both paths at the same time. Then, upon encountering the second beam splitter it ends up *interfering with itself* and through wave interference cancels out the effects of the first beam splitter and returns back to the original state.

This is a textbook example of the so-called *wave-particle duality*, underpinning quantum mechanics. If this does not make sense, then the best we can do at this point

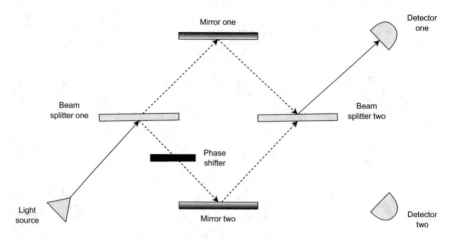

Fig. 2.5 Mach-Zehnder interferometer with light approaching from the bottom-left and a phase shifter

is to abandon our macroscopic, classical intuition and just trust the mathematics of quantum mechanics.

We can take this experiment further, and adjust the relative phase of the reflected light beams after the light goes through the first, but before the second beam splitter. This is illustrated by a phase shifter, placed in the path between beam splitter one and mirror two, as shown in Fig. 2.5.

The phase shifter allows changing the wavelength of the travelling light. Similarly to the basic modelling of the beam splitter we did earlier, we can also express the phase shifter as a unitary state transformation. Therefore, let us consider the matrix U_P, which performs the phase shift by ϕ.

$$U_P = \begin{bmatrix} 1 & 0 \\ 0 & e^{i\phi} \end{bmatrix} \tag{2.50}$$

We can now calculate the algebraic effect of this change. Since on Fig. 2.5 the input beam of light is approaching from the bottom-left, the state after the first beam splitter is the one from Eq. 2.47. This gives us the state after the phase shifter, but prior to beam splitter two that is equal to

$$U_P \left(\frac{1}{\sqrt{2}} (|u\rangle - |v\rangle) \right) = \frac{1}{\sqrt{2}} (|u\rangle - e^{i\phi} |v\rangle) \tag{2.51}$$

If we now imagine that the phase shifter in the Mach-Zehnder interferometer was set to π, we again get

$$e^{i\pi} = -1 \tag{2.52}$$

which, after substituting into Eq. 2.51 gives us

$$\frac{1}{\sqrt{2}}(|u\rangle - e^{i\phi}|v\rangle) = \frac{1}{\sqrt{2}}(|u\rangle + |v\rangle) \qquad (2.53)$$

Similarly as before, the extra phase shift by π manifested itself with a change in sign of the superposition terms—from $(-)$ to $(+)$. This has an interesting consequence that the state entering the second beam splitter corresponds no longer to the one from Eq. 2.47 but instead is the same as in Eq. 2.46. This, in turn means that the second beam splitter enacts the same transformation as in Eq. 2.48, namely

$$\mathbf{U_B}\left(\frac{1}{\sqrt{2}}(|u\rangle + |v\rangle)\right) = \begin{bmatrix} 1 \\ 0 \end{bmatrix} = |u\rangle \qquad (2.54)$$

which in our case is the opposite state from the one we started with—$|u\rangle$ instead of $|v\rangle$. The net effect is that the outcome is still deterministic, but adding a phase shifter allowed us to influence the character of the interference behavior, changing it from destructive to constructive interference. This changed the behavior of the detectors— as it forced detector one, associated with $|u\rangle$ to now fire in 100% of experimental runs. What is still troubling here, at least from the macroscopic standpoint, is that the effect manifests itself for both continuous light beam and individual photons, suggesting that manipulating a classical wave property such as wavelength, has an effect on the behavior on individual indivisible photons.

The exact same reasoning and calculation procedure could be applied to the initial light approaching from the top-left instead, in state $|u\rangle$. The conclusions would be identical to the one we managed to draw here, namely that the final state ends up being $|v\rangle$, which means the opposite detector than before is now detecting all the light.

All of the phenomena encountered in this section—the behavior of the beam splitter, the probabilities determining which of the detectors would fire, the interference exhibited via the addition of the second beam splitter, and finally the phase shifting allowing us to impact the character of interference will be critical components in our work with quantum computing from Chap. 4 onwards.

2.5   Observables

We have discussed the Born rule and how it embodies the probabilistic nature of quantum mechanics. We did not pay much attention to the theory of the measurement process itself, and, as it turns out, it is governed by its own set of rules that are much different to the acts of physical observation in classical physics.

Classically, a measurement of a system state, at least using idealized models, does not exert any significant influence on the system itself—any such external disturbance effects are negligible and normally do not need to be accounted for. Additionally, in classical physics the only limit to the measurement accuracy is the quality of the measurement technique and, in principle, a perfect measurement apparatus could deliver perfect results. Finally, there are no restrictions other than the complexity of the possible experimental setup in measuring different physical quantities at the same time.

The situation is dramatically different in quantum physics. Here, the measurement activities are destructive to the quantum state and simultaneous measurements of non-commuting properties are precluded. This process is governed by the concepts of projectors and observable-operators.

In Sect. 2.3 we defined the probabilities of measuring two possible states of the two level quantum system as

$$Pr(\uparrow) = |\langle \uparrow | \psi \rangle|^2 = |a|^2$$
$$Pr(\downarrow) = |\langle \downarrow | \psi \rangle|^2 = |b|^2 \tag{2.55}$$

We will now generalize this further, by observing that

$$|a|^2 = |\langle \uparrow | \psi \rangle|^2 = \langle \uparrow | \psi \rangle \langle \uparrow | \psi \rangle = \langle \psi | (| \uparrow \rangle \langle \uparrow |) | \psi \rangle \tag{2.56}$$

Because $|\uparrow\rangle \langle\uparrow|$ is an outer product of two vectors, it forms a matrix. Such matrix in quantum mechanical formalism is called a *projection operator*, or shorter *projector*. We can give such projector, which we shall refer to as $\mathbf{P}_\uparrow$, a more familiar matrix form by substituting the kets with vectors forming the Z-basis from Eq. 2.10

$$\mathbf{P}_\uparrow = |\uparrow\rangle \langle\uparrow| = \begin{bmatrix} 1 & 0 \\ 0 & 0 \end{bmatrix} \tag{2.57}$$

Similarly, we can use

$$|b|^2 = |\langle \downarrow | \psi \rangle|^2 = \langle \downarrow | \psi \rangle \langle \downarrow | \psi \rangle = \langle \psi | (| \downarrow \rangle \langle \downarrow |) | \psi \rangle \tag{2.58}$$

to calculate the projector for $\mathbf{P}_\downarrow$ as

$$\mathbf{P}_\downarrow = |\downarrow\rangle \langle\downarrow| = \begin{bmatrix} 0 & 0 \\ 0 & 1 \end{bmatrix} \tag{2.59}$$

This gives us generalized form of Eq. 2.55 for the probabilities of obtaining the state $|\uparrow\rangle$ or $|\downarrow\rangle$ upon measurement

$$Pr(\uparrow) = \langle \psi | \mathbf{P}_\uparrow | \psi \rangle$$
$$Pr(\downarrow) = \langle \psi | \mathbf{P}_\downarrow | \psi \rangle \tag{2.60}$$

Our projectors act on a quantum state, a two level quantum state in this case, and retrieve the component of the arbitrary state $|\psi\rangle$ associated with either $|\uparrow\rangle$ or $|\downarrow\rangle$.

In linear algebra, an eigenvector[3] $|u\rangle$ of a matrix $\mathbf{M}$ is a vector that changes only by a scalar λ when the linear transformation represented by $\mathbf{M}$ is applied to it. That scalar value is then called the eigenvalue. In other words, a linear transformation $\mathbf{M}$ sends its eigenvector to multiples λ of itself.

$$\mathbf{M}|u\rangle = \lambda|u\rangle \tag{2.61}$$

In quantum mechanics all quantum state measurements are described by Hermitian observable-operators of size $n \times n$ such that

$$\mathbf{O} = \sum_{i=1}^{n} \lambda_i \mathbf{P_i} \tag{2.62}$$

There are two particular characteristics of Hermitian operators that make them suitable for usage in quantum measurements. First of all, all eigenvalues of such an operator are guaranteed to be real numbers. Secondly, the eigenvectors associated with the distinct eigenvalues are orthogonal—which means they are always distinguishable and correspond to the notion of a basis.

We can now substitute Eqs. 2.57 and 2.59 into 2.62 to obtain the definition of an observable used for performing a quantum measurement using the Z-basis:

$$\mathbf{O_z} = \lambda_1 \mathbf{P_\uparrow} + \lambda_2 \mathbf{P_\downarrow} = \lambda_1 |\uparrow\rangle\langle\uparrow| + \lambda_2 |\downarrow\rangle\langle\downarrow| \tag{2.63}$$

The values of λ_i can be arbitrarily set to anything, as long as they are distinct from each other. Let us continue with the example of electron spin, which, as we remarked earlier, can take values $+\frac{\hbar}{2}$ for spin up, which we will use as λ_1 and $-\frac{\hbar}{2}$ for spin down, which we will set as λ_2. This gives us an observable $\mathbf{S_z}$ for measuring the spin component along the z-axis

$$\mathbf{S_z} = \frac{\hbar}{2}|\uparrow\rangle\langle\uparrow| - \frac{\hbar}{2}|\downarrow\rangle\langle\downarrow| = \frac{\hbar}{2}\begin{bmatrix} 1 & 0 \\ 0 & -1 \end{bmatrix} \tag{2.64}$$

It is worth noting that observable-operators in quantum mechanics do not directly act on the quantum state vector. Instead it is the relevant projector $\mathbf{P_i}$ associated with that observable, which transforms the quantum state vector into an eigenstate of the observable. This eigenstate has a distinct eigenvalue λ_i associated with it, representing the experimental outcome, which can be used i.e. to calculate statistical average of a repeated set of experiments.

All of this has a profound consequence, because it also means that in quantum mechanics a measurement is a destructive, non-reversible process—after measure-

---

[3] In generic linear algebra it is common to use the term *eigenvector*, while when referring to physical states such as quantum states, a more common term is *eigenstate*.

ment, regardless of the earlier state of the quantum object, it will always find itself in the same eigenstate of the observable associated with one specific λ_i eigenvalue.

Observables are responsible for a very unusual aspect of quantum mechanics, namely, the fact that observable quantities of the quantum system are not individual variables but rather operators. This is true for all physical measurable properties of any quantum system. A somewhat helpful way to think about such state of affairs might be to remind oneself that the quantum state is an abstract n-dimensional mathematical entity only, and in order to extract a specific physical measurable quantity out of it, we need to perform a "conversion"—and that is what we have observables for.

Even though a quantum system, in our example a simple two-level electron spin along the z-axis $|\psi\rangle$, can be described by and evolve along a continuous range of complex probability amplitudes a and b, only two possible discrete real numerical values can be extracted out of it—corresponding to the eigenvalues of the observable-operator. This, of course, also aligns with the physical reality we experience—as experimentally the spin indeed can take only ever yield one of two possible values $+\frac{\hbar}{2}$ or $-\frac{\hbar}{2}$.

In the earlier part of this chapter we discussed that quantum state is described using an orthonormal basis, and that any that we can always find infinitely many orthogonal unit vectors sets to form those. We can now elaborate on this further, by saying that it is the choice of the observable-operator that determines the measurement basis (eigenstates) and the range of possible measurement results (eigenvalues).

If we omit[4] the $\frac{\hbar}{2}$ from the matrix definition (2.64), we arrive at:

$$\sigma_z = \begin{bmatrix} 1 & 0 \\ 0 & -1 \end{bmatrix} \tag{2.65}$$

This is one of the so-called Pauli matrices, or Pauli operators, which are of fundamental importance in quantum physics and represent observables in three spatial directions—along the z-axis, x-axis and y-axis. We calculated the σ_z matrix using the projectors constructed around the quantum states from Eq. 2.10. In a similar fashion, using the vectors from Eqs. 2.17 (X-basis) and 2.22 (Y-basis), we could arrive at Pauli matrices σ_x and σ_y. The fourth Pauli matrix is referred to as an identity operator and is really the identity matrix from Eq. 2.33.

$$\sigma_x = \begin{bmatrix} 0 & 1 \\ 1 & 0 \end{bmatrix} \quad \sigma_y = \begin{bmatrix} 0 & -i \\ i & 0 \end{bmatrix} \quad \sigma_z = \begin{bmatrix} 1 & 0 \\ 0 & -1 \end{bmatrix} \quad \sigma_i = \begin{bmatrix} 1 & 0 \\ 0 & 1 \end{bmatrix} \tag{2.66}$$

Another consequence of having observable-operators representing the process of measurement, is something that Heisenberg realized when he discovered quantum mechanics—different measurements do not generally commute. This means that the process of extraction of real values out of the abstract quantum state depends on the order of measurements.

---

[4] Technically we use ± 1 instead of $\pm \frac{\hbar}{2}$ to set the values of λ_1 and λ_2.

In mathematics, we speak of an operation on two properties as being commutative when changing the order of operands produces the same result. Most obviously, addition and multiplication of real numbers is commutative, while division and subtraction of real numbers does not commute. In linear algebra, we define the commutator for two operators **A** and **B** as

$$[\mathbf{A}, \mathbf{B}] = \mathbf{A}\mathbf{B} - \mathbf{B}\mathbf{A} \tag{2.67}$$

The operators commute when

$$[\mathbf{A}, \mathbf{B}] = 0 \tag{2.68}$$

and do not commute when

$$[\mathbf{A}, \mathbf{B}] \neq 0 \tag{2.69}$$

We could calculate this for the Pauli matrices, for example for σ_x and σ_z:

$$\sigma_x \sigma_z = \begin{bmatrix} 0 & 1 \\ 1 & 0 \end{bmatrix} \begin{bmatrix} 1 & 0 \\ 0 & -1 \end{bmatrix} = \begin{bmatrix} 0 & -1 \\ 1 & 0 \end{bmatrix} = -i\sigma_y \tag{2.70}$$

$$\sigma_z \sigma_x = \begin{bmatrix} 1 & 0 \\ 0 & -1 \end{bmatrix} \begin{bmatrix} 0 & 1 \\ 1 & 0 \end{bmatrix} = \begin{bmatrix} 0 & 1 \\ -1 & 0 \end{bmatrix} = i\sigma_y \tag{2.71}$$

The first thing that jumps out here, is that there is a relation between Pauli matrices—a product of σ_x and σ_z, produced σ_y multiplied by i or $-i$. Secondly, we notice that the multiplication result of the matrices is different depending on the order in which we multiply them. As a result:

$$[\sigma_x, \sigma_z] = -i\sigma_y - i\sigma_y = -2i\sigma_y \tag{2.72}$$

We can say with certainty here, that Pauli X and Z operators do not commute, and hence we cannot measure a property of a quantum object, for example the spin of an electron, along the x-axis (using σ_x) and then along the z-axis (using σ_z), and expect to obtain the same results as measuring in a reverse order.

A different way of looking at it, is to consider the eigenvectors of the Pauli matrices. Two operators commute only if they have the same eigenvectors—and this is clearly not the case here.

$$\sigma_x: \quad |\leftarrow\rangle = \frac{1}{\sqrt{2}} \begin{bmatrix} 1 \\ 1 \end{bmatrix} \quad |\rightarrow\rangle = \frac{1}{\sqrt{2}} \begin{bmatrix} 1 \\ -1 \end{bmatrix} \tag{2.73}$$

$$\sigma_y: \quad |\circlearrowright\rangle = \frac{1}{\sqrt{2}} \begin{bmatrix} 1 \\ i \end{bmatrix} \quad |\circlearrowleft\rangle = \frac{1}{\sqrt{2}} \begin{bmatrix} 1 \\ -i \end{bmatrix} \tag{2.74}$$

$$\sigma_z: \qquad |\uparrow\rangle = \begin{bmatrix} 1 \\ 0 \end{bmatrix} \qquad |\downarrow\rangle = \begin{bmatrix} 0 \\ 1 \end{bmatrix} \tag{2.75}$$

Since after the measurement using a given observable-operator the quantum state collapses to the eigenstate corresponding to that observable, it is clear that if the eigenvectors are different between the two operators, then those measurements are not compatible with each other. For example, a measurement using σ_z would leave the quantum state in one of the eigenstates $|\uparrow\rangle$ or $|\downarrow\rangle$. A subsequent σ_x measurement would then put that system into the eigenstate $|\leftarrow\rangle$ or $|\rightarrow\rangle$. Finally, a follow up measurement using σ_z again, is now not guaranteed to yield the same result as initially—because the incompatible measurement σ_x disturbed the system in a way that created maximum uncertainty of the measurement outcome.

This is the core idea behind the famous *Heisenberg uncertainty principle*, which tells us that we cannot simultaneously know values of two physical properties of a quantum system represented by incompatible observables with maximum confidence.

2.6 Larger Systems

2.6.1 Tensor Product

So far we only considered simple two-level systems, such as a spin of a single electron. The situation gets even more interesting once we start moving to more complex composite systems, starting already with trying to describe two particles. Quantum mechanics uses the algebraic concept of *tensor product* of the vector spaces to provide abstract description of such systems. Such description is a consequence of the formalism based on Hilbert spaces, because the Hilbert space of a compound quantum system is the tensor product of the Hilbert spaces of the component subsystems [12].

$$\mathcal{H}^{(N)} = \mathbb{C}^2 \otimes \mathbb{C}^2 \otimes \dots \otimes \mathbb{C}^2 \tag{2.76}$$

Let us have two simple two-level quantum systems, $|\psi\rangle$ and $|\varphi\rangle$. As we already know, the quantum state of each one can be independently described by a two dimensional complex vector:

$$|\psi\rangle = \begin{bmatrix} a_1 \\ b_1 \end{bmatrix} \qquad |\varphi\rangle = \begin{bmatrix} a_2 \\ b_2 \end{bmatrix} \tag{2.77}$$

This can of course be rewritten into a more digestible linear combination—superposition—of two basis states, taking a Z-basis $\{|\uparrow\rangle, |\downarrow\rangle\}$ basis from Eq. 2.10 as a reference point.

$$|\psi\rangle = a_1 |\uparrow\rangle + b_1 |\downarrow\rangle \qquad |\varphi\rangle = a_2 |\uparrow\rangle + b_2 |\downarrow\rangle \tag{2.78}$$

A tensor product between two vectors belonging to $\mathbb{C}^2$, a two-dimensional space, creates a single vector in $\mathbb{C}^4$, a four-dimensional space

$$\begin{bmatrix} a_1 \\ b_1 \end{bmatrix} \otimes \begin{bmatrix} a_2 \\ b_2 \end{bmatrix} = \begin{bmatrix} a_1 a_2 \\ a_1 b_2 \\ b_1 a_2 \\ b_1 b_2 \end{bmatrix} \tag{2.79}$$

Just like a two-dimensional state vector can be represented using a two state orthonormal basis, a four state orthonormal basis is relevant for the composite system consisting of two two-level systems. Therefore, in our example, the orthonormal basis for the tensor product $|\psi\rangle \otimes |\varphi\rangle$ is

$$\{|\uparrow\rangle |\uparrow\rangle, |\uparrow\rangle |\downarrow\rangle, |\downarrow\rangle |\uparrow\rangle, |\downarrow\rangle |\downarrow\rangle\} \tag{2.80}$$

In the Dirac braket notation, in order to make things more succinct and readable, it is customary to merge the neighboring kets into a single ket. As such, we can also express the four dimensional basis as:

$$\{|\uparrow\uparrow\rangle, |\uparrow\downarrow\rangle, |\downarrow\uparrow\rangle, |\downarrow\downarrow\rangle\} \tag{2.81}$$

We can denote the composite state as $|\phi\rangle$

$$|\phi\rangle = |\psi\rangle \otimes |\varphi\rangle = a_1 a_2 |\uparrow\uparrow\rangle + a_1 b_2 |\uparrow\downarrow\rangle + b_1 a_2 |\downarrow\uparrow\rangle + b_1 b_2 |\downarrow\downarrow\rangle \tag{2.82}$$

The probability of obtaining any of the four possible outcomes is still governed by the Born rule

$$Pr(\uparrow\uparrow) = |\langle\uparrow\uparrow |\phi\rangle|^2 = |a_1 a_2|^2$$
$$Pr(\uparrow\downarrow) = |\langle\uparrow\downarrow |\phi\rangle|^2 = |a_1 b_2|^2$$
$$Pr(\downarrow\uparrow) = |\langle\downarrow\uparrow |\phi\rangle|^2 = |b_1 a_2|^2$$
$$Pr(\downarrow\downarrow) = |\langle\downarrow\downarrow |\phi\rangle|^2 = |b_1 b_2|^2 \tag{2.83}$$

One final thing worth noting, is that in order to provide more concise way of expressing abstract quantum states, especially those that are large, it is customary in the Dirac notation to omit the $\otimes$ symbol between the kets that are subject to the the tensor product and collapse them into a single ket.

$$|\psi\rangle \otimes |\varphi\rangle = |\psi\varphi\rangle = |\phi\rangle \tag{2.84}$$

We could extend all of this reasoning to a composite system consisting of three quantum subsystems, at which point an obvious pattern should start to emerge. A tensor product between three vectors belonging to $\mathbb{C}^2$ creates a single vector in $\mathbb{C}^8$— an eight-dimensional vector space.

$$\begin{bmatrix} a_1 \\ b_1 \end{bmatrix} \otimes \begin{bmatrix} a_2 \\ b_2 \end{bmatrix} \otimes \begin{bmatrix} a_3 \\ b_3 \end{bmatrix} = \begin{bmatrix} a_1 a_2 a_3 \\ a_1 a_2 b_3 \\ a_1 b_2 a_3 \\ \vdots \\ b_1 b_2 b_3 \end{bmatrix} \tag{2.85}$$

The orthonormal basis consists of eight state vectors as well, representing all the possible basis state arrangements between the subsystems. The tensor product qualification is an important one, because that is what gives rise to this large number of dimensions. We can generalize it by saying that the composite quantum system is described by a single 2^n dimensional abstract state vector, in a $\mathbb{C}^n$ vector space, where n is the number of constituent two-level subsystems.

$$|\psi\rangle = |\psi_1\rangle \otimes |\psi_2\rangle \ldots \otimes |\psi_n\rangle \tag{2.86}$$

This growth of complexity of the system is exponential, and therefore by the time we reach $n = 260$—which by all means is still a rather simple quantum system—the composite quantum state is described by $2^{260} \approx 10^{80}$ complex numbers—each contributing an extra dimension. 10^{80} happens to be roughly the estimated amount of atoms in the observable universe.[5] To echo Carlton M. Caves [3]—Hilbert space is a big place indeed.

2.6.2 Entanglement

In quantum mechanics, it is also possible to have composite states that are non-factorizable to the tensor product of the individual subsystems components. An example of such state is

$$|\psi\rangle = \frac{1}{\sqrt{2}}\left(|\uparrow\downarrow\rangle + |\downarrow\uparrow\rangle\right) \tag{2.87}$$

In what is rather surprising state of affairs, $|\psi\rangle$ above, is a perfectly valid description of a two particle system, even though it is incompatible with Eq. 2.82. That is of course because we could write it as

$$|\psi\rangle = 0|\uparrow\uparrow\rangle + \frac{1}{\sqrt{2}}|\uparrow\downarrow\rangle + \frac{1}{\sqrt{2}}|\downarrow\uparrow\rangle + 0|\downarrow\downarrow\rangle \tag{2.88}$$

---

[5] This is the so-called "Eddington Number", derived initially by Arthur Eddington [11]. At the same time, in these examples we are still in a finite-dimensional context—but in quantum mechanics in general, in case of continuous variables, the mathematical structure can actually be infinite-dimensional too.

This gives us

$$a_1a_2 = 0 \quad a_1b_2 = \frac{1}{\sqrt{2}} \quad b_1a_2 = \frac{1}{\sqrt{2}} \quad b_1b_2 = 0 \tag{2.89}$$

From that we can easily see that we cannot find a_1, a_2, b_1 and b_2 that would satisfy the decomposition into component states.

The quantum state (2.87) is an example of an EPR pair. It resembles a uniform superposition state from Eq. 2.16, however instead of one two-level quantum system, it encompasses two of them—for example two photons. The individual particles can still be measured independently from each other. During measurement, the Born rule applies and tells us that the particles are perfectly correlated.

When one of the constituents of the system is measured, it will collapse to a state $|\uparrow\rangle$ or $|\downarrow\rangle$ with equal probability and the other one will then be guaranteed to be measured to the opposite state. Such particles could even be measured at the exact same moment in time, precluding any chance of them "communicating"—they still exhibit this perfect correlation between each other as predicted by quantum mechanics. This phenomenon is called *entanglement* and is a distinctly quantum feature, with no counterpart effect in classical physics. The primary thing the formalism of quantum mechanics tells us about such entangled quantum objects, is that they are impossible to be treated independently anymore, instead, despite the fact they can be spatially separated, they can only be reasoned about as a single state.

Contrary to superposition, entanglement is not basis dependent—it is not possible for quantum objects to be entangled in one basis, and not be entangled in another. However the recorded correlations of measurement results depend on the basis choice of each of the two observers.

The state (2.87) is one of the four so-called *Bell states*, which represent four different ways of having maximal entanglement between pairs of two-level quantum objects. Together, they form an orthonormal basis for such system:

$$|\Phi^+\rangle = \frac{1}{\sqrt{2}}\left(|\uparrow\uparrow\rangle + |\downarrow\downarrow\rangle\right)$$

$$|\Phi^-\rangle = \frac{1}{\sqrt{2}}\left(|\uparrow\uparrow\rangle - |\downarrow\downarrow\rangle\right)$$

$$|\Psi^+\rangle = \frac{1}{\sqrt{2}}\left(|\uparrow\downarrow\rangle + |\downarrow\uparrow\rangle\right)$$

$$|\Psi^-\rangle = \frac{1}{\sqrt{2}}\left(|\uparrow\downarrow\rangle - |\downarrow\uparrow\rangle\right) \tag{2.90}$$

The states $|\Phi^+\rangle$ and $|\Phi^-\rangle$ are positively correlated—measurement of a given result on one of the particles making up the pair, guarantees that the second one will have the produce the same result as well. On the other hand, the states $|\Psi^+\rangle$ and $|\Psi^-\rangle$ are negatively correlated—once one part of the pair is measured and produces a certain result, the other is certain to produce an opposite measurement outcome.

2.7 No-Cloning Theorem

The mathematical foundations of quantum mechanics, or more specifically the unitary character of all quantum state transformations, preclude the possibility of being able to copy an unknown quantum state in what is known as *no-cloning theorem*.

   The theorem can be proven using negative proof (proof of impossibility). Consider a hypothetical universal two-state copying unitary transformation **C**. If it existed, being universal, it would be reasonable to expect that an arbitrary unknown state $|\varphi\rangle$ and another arbitrary unknown state $|\psi\rangle$ can be copied onto a a blank state $|\Omega\rangle$, such that:

$$\mathbf{C}(|\varphi\rangle \otimes |\Omega\rangle) = |\varphi\rangle \otimes |\varphi\rangle$$
$$\mathbf{C}(|\psi\rangle \otimes |\Omega\rangle) = |\psi\rangle \otimes |\psi\rangle \tag{2.91}$$

   Let us now consider a third state, $|\phi\rangle$, which is a linear combination both $|\varphi\rangle$ and $|\psi\rangle$. A general definition of $|\phi\rangle$ would then be as defined in Eq. 2.92.

$$|\phi\rangle = \frac{1}{\sqrt{2}}\left(a\,|\varphi\rangle + b\,|\psi\rangle\right) \tag{2.92}$$

   We could now combine (2.91) and (2.92) in order to, still reasonably, expect that our copying **C** transformation is equally capable of copying $|\phi\rangle$ onto the scratch pure state $|\Omega\rangle$ as well. The expected result would then be

$$|\phi\rangle \otimes |\phi\rangle = \frac{1}{2}\left(a^2\,|\varphi\varphi\rangle + ab\,|\varphi\psi\rangle + ab\,|\psi\varphi\rangle + b^2\,|\psi\psi\rangle\right) \tag{2.93}$$

   At this point things seem to be going logically well, but we are about to hit a dead end. Because of linearity of the transformations, we could actually write the copying effect of **C** on the $|\phi\rangle$ state from Eq. 2.92 as:

$$\mathbf{C}(|\phi\rangle \otimes |\Omega\rangle) = \frac{1}{\sqrt{2}}\left(\mathbf{C}(|\varphi\rangle \otimes |\Omega\rangle) + \mathbf{C}(|\psi\rangle \otimes |\Omega\rangle)\right) = \frac{1}{\sqrt{2}}\left(|\varphi\varphi\rangle + |\psi\psi\rangle\right) \tag{2.94}$$

   Of course the two states (2.93) and (2.94) contradict themselves, which in turn means that **C** with its universal copying properties cannot exist.

   There is also a different angle we take here, namely to inspect the inner products of the inputs and the outputs of **C**. Recall that all unitary transformations preserve inner products. This means that our copying transformation **C** would have to be bound by these rules as well. Consider again the copying expectations defined in Eq. 2.91 for two unknown states $|\varphi\rangle$ and $|\psi\rangle$. The inner product of the inputs into **C** is:

$$\langle \Omega\,\psi | \varphi\,\Omega \rangle = \langle \Omega | \Omega \rangle \langle \psi | \varphi \rangle = \langle \psi | \varphi \rangle \tag{2.95}$$

The product of the outputs of $\mathbf{C}$ is:

$$\langle \psi\psi | \varphi\varphi \rangle = \langle \psi | \varphi \rangle \langle \psi | \varphi \rangle = \langle \psi | \varphi \rangle^2 \tag{2.96}$$

Because quantum mechanics requires $\mathbf{C}$ to preserve the inner product, and that we deal with unit vectors, whose length must normalize to 1, it follows that for a copying transformation $\mathbf{C}$:

$$\langle \psi | \varphi \rangle = \langle \psi | \varphi \rangle^2 \tag{2.97}$$

This creates an interesting constraint on copying. Equation 2.97 can only be satisfied when the product $\langle \psi | \varphi \rangle$ is either ± 1 or 0. In the first case, it would mean that the states $| \varphi \rangle$ and $| \psi \rangle$ are actually equivalent with each other $| \varphi \rangle = | \psi \rangle$, while the latter would indicate that the two states are linearly independent (orthogonal). This leads us to the conclusion that the only states a copying transformation $\mathbf{C}$ can copy are mutually orthogonal states. Moreover, we would still need a transformation specifically built for those states, as a universal copier cannot be constructed [9].

2.8 Postulates of Quantum Mechanics

Quantum mechanics is underpinned by a set of axioms—postulates that are taken to be true, and for which we do not have a deeper explanation *why* they are correct, other that they fit perfectly with all the empirical evidence. They have not been canonically codified, therefore various authors present them formulated differently, though the overall set of ideas they cover is consistent.

At this point we are equipped with enough foundational algebra-based quantum theory to be ready to state them—they should help organize the topics covered so far and establish a baseline for our understanding of quantum mechanics. In particular, Chuang and Nielsen [17] provide a formulation of the postulates that is geared towards application scenarios in quantum computing, which aligns very well with the theme of this book. The postulates of quantum mechanics presented here are therefore adapted from those.

1. Quantum system is completely described by a complex vector $| \psi \rangle$ of length 1 known as the *quantum state vector*, which exists in a complex vector space specific to this system, called the Hilbert space. Contrary to classical physics, the state does not describe the physical properties of the system but is an abstract entity that can be used to extract classical information from the system via the observables. Unknown quantum states cannot be copied.
2. Evolution of a closed system $| \psi \rangle$ happens by the application of unitary operator $\mathbf{U}$, described by the matrix whose size corresponds to the dimensions of the state vector. This condition also means that in a closed quantum system every transformation is reversible.

3. Observables are described by Hermitian matrices and used to mathematically express a measurement operation. The eigenvalues of these observables are real numbers and represent the possible values that can be observed, and therefore real physical properties of the system. The measurement outcome is probabilistic—for an observable $\mathbf{O}$ the system collapses to one of its eigenstates $|a\rangle$ with probability calculated according to the Born rule. The physical result of the measurement is the eigenvalue λ associated with the eigenstate $|a\rangle$. Measurement is non-unitary and irreversible.

4. A composite quantum system is described by a tensor product of the n component subsystems. The resulting state space is $\mathbb{C}^n$, and the state vector has 2^n dimensions.

When reading the postulates carefully, it quickly becomes apparent that the postulates contain a major consistency issue. The quantum state evolution happens according to unitary, reversible state transformations, which in essence means that the entire evolution is completely deterministic. That state, is however in principle unobservable, because the observations of quantum phenomena are governed by the rules related to observable-operators and the probabilistic state collapse according to the Born rule. This dichotomy between a deterministic and reversible system evolution and the probabilistic and irreversible measurement theory governed by observables, has become known as the *measurement problem*. It has baffled physicists and philosophers since the early days of quantum theory, leading to various interpretational frameworks, none of which has managed to solve the issue at hand with satisfiable and universally accepted way.

2.9 Bell's Theorem

2.9.1 Bell's Inequality

John S. Bell [1] imagined an experimental set up, based on David Bohm's EPR variant, in which two entangled particles in a singlet state $|\Psi^-\rangle$ fly off in the opposite directions, where they hit detectors, such as independent Stern-Gerlach spin measuring apparatuses, configured to be measuring the spin at angles θ_1 and θ_2. So far we only discussed measurement along the x-, y- or z-axis, but in principle a measurement along any spatial direction is possible, and those can be described using angles.

If $\theta_1 = \theta_2$, then both particles are measured in the same basis, so the original conditions from Bohm's EPR experiment are recovered and the recorded values are always opposite to each other. However, Bell realized that once the angles are different from each other, quantum mechanical correlations are not always reproducible using local realism.

Bell's inequality can be derived using minimal assumptions. Let us imagine three non-orthogonal vectors $\mathbf{a}$, $\mathbf{b}$ and $\mathbf{c}$ in a three-dimensional space, which provide directions for spin measurements. In each case, an observable $\mathbf{O_a}$, $\mathbf{O_b}$ and $\mathbf{O_c}$ associated with a given measurement direction can produce eigenvalues $+1$ or -1 as the result. Given that we deal with three directions, each one yielding two possible measurement outcomes, all possible permutations for the system measurements can be described using $2^3 = 8$ configuration sets.

We now consider Alice and Bob sharing EPR pairs in a singlet state $|\Psi^-\rangle$.

$$|\Psi^-\rangle = \frac{1}{\sqrt{2}}\left(|\uparrow\downarrow\rangle - |\downarrow\uparrow\rangle\right) \tag{2.98}$$

We know that if they measure their parts of the pair along the same direction, say **a**, the entanglement guarantees that they will receive the opposite measurement results. When they measure along different directions, for example Alice along **a** and Bob along **b** or **c** their measurements may end up being the same or come out with opposite values. The latter set of cases is what is interesting in deriving Bell's inequality.

Table 2.1 summarizes all possible measurement outcomes for measurements using observables $\mathbf{O_a}$, $\mathbf{O_b}$ and $\mathbf{O_c}$, into what we will refer to here as *groups*. It may look a little overwhelming at first, but is actually very easy to read. For example, group N_4 describes a case where:

- if Alice measures along the **a** direction, she would receive $+1$. Bob would then receive -1 for the **a** direction, $+1$ for the **b** direction and $+1$ for the **c** direction measurement.
- if Alice measures along the **b** direction, she would receive -1. Bob would still receive -1 for the **a** direction, $+1$ for the **b** direction and $+1$ for the **c** direction measurement.
- if Alice measures along the **c** direction, she would receive -1. Yet again, Bob would still receive -1 for the **a** direction, $+1$ for the **b** direction and $+1$ for the **c** direction measurement.

Notice that we can clearly see here that, as mentioned before, whenever Alice and Bob measure in the same direction, their results are always correlated in the opposite way, while for the situations where the measurement directions do not agree, the

Table 2.1 Possible EPR pair measurement results for Alice and Bob in directions **a**, **b** and **c**

	Alice particle	Alice measurement			Bob particle	Bob measurement		
		a	b	c		a	b	c
N_1	(a^+, b^+, c^+)	$+1$	$+1$	$+1$	(a^-, b^-, c^-)	-1	-1	-1
N_2	(a^+, b^+, c^-)	$+1$	$+1$	-1	(a^-, b^-, c^+)	-1	-1	$+1$
N_3	(a^+, b^-, c^+)	$+1$	-1	$+1$	(a^-, b^+, c^-)	-1	$+1$	-1
N_4	(a^+, b^-, c^-)	$+1$	-1	-1	(a^-, b^+, c^+)	-1	$+1$	$+1$
N_5	(a^-, b^+, c^+)	-1	$+1$	$+1$	(a^+, b^-, c^-)	$+1$	-1	-1
N_6	(a^-, b^+, c^-)	-1	$+1$	-1	(a^+, b^-, c^+)	$+1$	-1	$+1$
N_7	(a^-, b^-, c^+)	-1	-1	$+1$	(a^+, b^+, c^-)	$+1$	$+1$	-1
N_8	(a^-, b^-, c^-)	-1	-1	-1	(a^+, b^+, c^+)	$+1$	$+1$	$+1$

results appear random. We can use the methodology described here to study all the other cases that could possibly occur.

One particular aspect worth paying attention to is how many possible measurement configurations are there in each group if we require that Alice and Bob would use different measurement direction. The answer to this is six: Alice measures along $\mathbf{a}$ and Bob measures along $\mathbf{b}$ or $\mathbf{c}$, Alice measures along $\mathbf{b}$ and Bob measures along $\mathbf{a}$ or $\mathbf{c}$ and finally, Alice measures along $\mathbf{c}$ and Bob measures along $\mathbf{a}$ or $\mathbf{b}$.

A different way of reading the table is to ask a question such as "if Alice measures $+1$ along the $\mathbf{a}$ direction, how many times Bob can measure $+1$ along the $\mathbf{b}$ direction"? Careful study of the possibilities leads us to the conclusion that this can only happen twice out of the eight variants, and is described in groups N_3 and N_4. We can of course assign a probability for Alice's and Bob's measurements to end up in those groups N_3 and N_4, and that is:

$$Pr(\mathbf{a^+}, \mathbf{b^+}) = \frac{N_3 + N_4}{N} \tag{2.99}$$

where

$$N = N_1 + N_2 + N_3 + N_4 + N_5 + N_6 + N_7 + N_8 \tag{2.100}$$

Using similar reasoning, we could also calculate $Pr(\mathbf{b^+}, \mathbf{c^+})$. These are cases where Alice measures along $+1$ in direction $\mathbf{b}$, while Bob measures $+1$ in direction $\mathbf{b}$, and they only occur for groups N_2 and N_6

$$Pr(\mathbf{b^+}, \mathbf{c^+}) = \frac{N_2 + N_6}{N} \tag{2.101}$$

Finally, $Pr(\mathbf{a^+}, \mathbf{c^+})$ can be obtained by looking at groups N_2 and N_6

$$Pr(\mathbf{a^+}, \mathbf{c^+}) = \frac{N_2 + N_4}{N} \tag{2.102}$$

Given that these are probabilities, we know that they are real numbers, and that they are all larger than or equal to 0. We could therefore confidently say that their sum will also be larger than 0.

$$Pr(\mathbf{a^+}, \mathbf{b^+}) + Pr(\mathbf{b^+}, \mathbf{c^+}) + Pr(\mathbf{a^+}, \mathbf{c^+}) \geq 0$$
$$\frac{N_3 + N_4 + N_2 + N_6 + N_2 + N_4}{N} \geq 0 \tag{2.103}$$

If we examine Eqs. 2.99, 2.101 and 2.102 slowly, we can observe that both N_2 and N_4 appear twice in them. We could therefore reduce them by subtracting $Pr(\mathbf{a^+}, \mathbf{c^+})$ instead of adding it

$$Pr(\mathbf{a^+}, \mathbf{b^+}) + Pr(\mathbf{b^+}, \mathbf{c^+}) - Pr(\mathbf{a^+}, \mathbf{c^+}) \geq 0 \qquad (2.104)$$

The changed variant will still hold because we can expand it to

$$\frac{N_3 + N_4 + N_2 + N_6 - N_2 - N_4}{N} \geq 0$$

$$\frac{N_3 + N_6}{N} \geq 0 \qquad (2.105)$$

so what we are doing are still just adding real positive numbers to each other, and their sum will always be greater than or equal to 0. This leads us to the final formulation of Bell's inequality

$$Pr(\mathbf{a^+}, \mathbf{b^+}) + Pr(\mathbf{b^+}, \mathbf{c^+}) \geq Pr(\mathbf{a^+}, \mathbf{c^+}) \qquad (2.106)$$

Bell's inequality from Eq. 2.106 can be used to test local realistic (hidden variables) theories—they will obey it, while quantum mechanics will not. That is the core idea behind Bell's theorem, which states that:

No physical theory of local hidden variables can ever reproduce all of the predictions of quantum mechanics.

Given that superposition is basis independent, and by relying on the rotational symmetry, we can express the singlet state $|\Psi^-\rangle$ with reference to any arbitrary axis, such as our three directional vectors $\mathbf{a}$, $\mathbf{b}$ or $\mathbf{c}$. Let us assume that $|u\rangle$ and $|v\rangle$ are the eigenvectors[6] of the observable σ in each of the directions—we will designate the eigenstate $|u\rangle$ as corresponding to the result $+1$ and $|v\rangle$ to -1. Based on the singlet state $|\Psi^-\rangle$ we can then write

$$|\psi\rangle = \frac{1}{\sqrt{2}}\left(|u_a\rangle |v_a\rangle - |v_a\rangle |u_a\rangle\right) =$$
$$\frac{1}{\sqrt{2}}\left(|u_b\rangle |v_b\rangle - |v_b\rangle |u_b\rangle\right) =$$
$$\frac{1}{\sqrt{2}}\left(|u_c\rangle |v_c\rangle - |v_c\rangle |u_c\rangle\right) = \qquad (2.107)$$

With such singlet state representations, we can relate all three of them together and calculate probabilities for any of the two axis pairs, with the relevant θ angle between them. For example for $Pr(\mathbf{a^+}, \mathbf{b^+})$ we would take the inner product $\langle u_a u_b | \psi \rangle$. This gives us

---

[6] We could have used any labels here, but it is better to avoid using arrows as so far they implied for us specific spatial directions: z-axis was noted with $\{|\uparrow\rangle, |\downarrow\rangle\}$, x-axis with $\{|\leftarrow\rangle, |\rightarrow\rangle\}$, while y-axis with $\{|\circlearrowleft\rangle, |\circlearrowright\rangle\}$. In this case we deal with parameterized directions so generic labels are more appropriate.

$$\langle u_a u_b | \psi \rangle = \frac{1}{\sqrt{2}} \left(\langle u_a | u_a \rangle \langle u_b | v_a \rangle - \langle u_a | v_a \rangle \langle u_b | u_a \rangle \right) = \frac{1}{\sqrt{2}} \langle u_b | v_a \rangle \qquad (2.108)$$

In Eq. 2.24 we defined a general approach for reasoning about a two-level quantum state using a real-space three-dimensional spherical representation. Using that we can express $\langle u_b | v_a \rangle$ as:

$$\frac{1}{\sqrt{2}} \langle u_b | v_a \rangle = \frac{1}{\sqrt{2}} \sin \left(\frac{\theta_{ab}}{2} \right) \qquad (2.109)$$

This is of course still the probability amplitude, and the actual probability $Pr(\mathbf{a^+}, \mathbf{b^+})$ will be

$$Pr(\mathbf{a^+}, \mathbf{b^+}) = \left| \frac{1}{\sqrt{2}} \sin \left(\frac{\theta_{ab}}{2} \right) \right|^2 = \frac{1}{2} \sin^2 \left(\frac{\theta_{ab}}{2} \right) \qquad (2.110)$$

Based on this exact procedure, $Pr(\mathbf{a^+}, \mathbf{c^+})$ and $Pr(\mathbf{b^+}, \mathbf{c^+})$ would be

$$Pr(\mathbf{a^+}, \mathbf{c^+}) = \frac{1}{2} \sin^2 \left(\frac{\theta_{ac}}{2} \right)$$

$$Pr(\mathbf{b^+}, \mathbf{c^+}) = \frac{1}{2} \sin^2 \left(\frac{\theta_{bc}}{2} \right) \qquad (2.111)$$

By substituting those into Eq. 2.106 they together give us the Bell's inequality

$$\frac{1}{2} \sin^2 \left(\frac{\theta_{ab}}{2} \right) + \frac{1}{2} \sin^2 \left(\frac{\theta_{bc}}{2} \right) \geq \frac{1}{2} \sin^2 \left(\frac{\theta_{ac}}{2} \right) \qquad (2.112)$$

Furthermore, if we select $\mathbf{a}$, $\mathbf{b}$ or $\mathbf{c}$ such that they are anti-clockwise about the z-axis, we can set up the angles such that

$$\theta_{ab} = \theta_{bc} \qquad (2.113)$$
$$\theta_{ac} = \theta_{ab} + \theta_{bc}$$

With that, we can now combine Eqs. 2.112 and 2.113 to reduce the inequality to just a single angle θ. In the end, this gives us the final simplified Bell's inequality

$$\sin^2 \left(\frac{\theta}{2} \right) \geq \frac{1}{2} \sin^2 (\theta) \qquad (2.114)$$

Let us now imagine that we set θ as shown in Fig. 2.6

$$\theta = \frac{\pi}{3} = 60° \qquad (2.115)$$

Fig. 2.6 Bell inequality violation vectors with $\frac{\pi}{3}$ angle between each neighboring pair of vectors

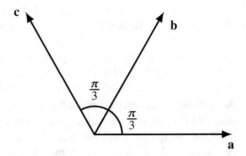

Indeed, the inequality (2.114) evaluates then to a result that is a clear violation

$$\frac{1}{4} \ngeq \frac{3}{8} \tag{2.116}$$

2.9.2 CHSH Inequality

The CHSH inequality [6] uses similar principles that Bell relied on and is based on two orthogonal vector pairs $\mathbf{a}$, $\mathbf{a}'$, $\mathbf{b}$ and $\mathbf{b}'$.

To understand the CHSH angle of approaching the problem, we need to imagine the following setup. Alice and Bob share EPR pairs and can measure their parts independently. Alice uses measurement apparatus configured with two possible observables $\mathbf{O_a}$ and $\mathbf{O_{a'}}$, in order to measure spin along $\mathbf{a}$ and $\mathbf{a}'$ directions respectively. We will refer to her possible measurement results as A and A'. Bob uses measurement setups $\mathbf{O_b}$ and $\mathbf{O_{b'}}$, corresponding to directions $\mathbf{b}$ and $\mathbf{b}'$, and his possible results are B and B'. The possible measurement outcomes, so eigenvalues of the observables used and values of A, A', B and B' are of course in all cases still equal to ± 1.

Here is where the locality hypothesis comes into play. We assume, based on locality, that values of measurement outcomes are predetermined by some hidden variables set γ shared between the EPR pair. Consequently, we assume that Alice's and Bob's measurement processes do not influence each other—for example, B does not depend on the outcome of A or A' and is instead entirely conditioned on the local apparatus setup and hidden variables γ only. With that in mind, we can write the products of the measurement outcomes after [10] as:

$$S = AB + AB' + A'B - A'B' = A(B + B') + A'(B - B') \tag{2.117}$$

Because all values can only take values of either $+1$ or -1, one of the expressions—$A(B + B')$ or $A'(B - B')$ will always evaluate to zero, leaving us with $S = 2$ or $S = -2$. From this, we can say that a statistical average of $\langle S \rangle$ must be within this range too

$$-2 \leq \langle S \rangle \leq 2 \tag{2.118}$$

This is the CHSH inequality, which is satisfied by any average values data set produced by independent measurement processes of correlated data, as long as their results are in ± 1 form, and—of course!—in which locality is obeyed [14]. In other words, processes in which the respective measurement outcomes, performed by separate observers, do not influence themselves.

The quantum mechanical variant of CHSH inequality can be obtained by re-using our measurement outcome probabilities for the singlet state, defined as functions of angle θ between the two vectors, defined in Eqs. 2.110 and 2.111. From that, we can construct the quantum mechanical correlation function, which is a de facto expectation value of the product of possible measurement outcomes. More specifically, the probabilities associated with specific eigenvalue outcomes for the EPR pair, can be multiplied by the product of those eigenvalues, to help us construct the expectation value.

For example, for measurement directions $\mathbf{a}$ and $\mathbf{b}$, the eigenvalue products are:

- $(+1)(+1) = 1$ for $(\mathbf{a^+}, \mathbf{b^+})$
- $(+1)(-1) = -1$ for $(\mathbf{a^+}, \mathbf{b^-})$
- $(-1)(+1) = -1$ for $(\mathbf{a^-}, \mathbf{b^+})$
- $(-1)(-1) = 1$ for $(\mathbf{a^-}, \mathbf{b^-})$

which then produces the specific expectation value for $E(\mathbf{a}, \mathbf{b})$:

$$E(\mathbf{a}, \mathbf{b}) = Pr(\mathbf{a^+}, \mathbf{b^+}) + Pr(\mathbf{a^-}, \mathbf{b^-}) - Pr(\mathbf{a^+}, \mathbf{b^-}) - Pr(\mathbf{a^-}, \mathbf{b^+}) =$$

$$\sin^2\left(\frac{\theta}{2}\right) - \cos^2\left(\frac{\theta}{2}\right) = -cos(\theta) \qquad (2.119)$$

Repeating the same procedure for the other vector pairs, we can obtain expectation values $E(\mathbf{a}, \mathbf{b'})$, $E(\mathbf{a'}, \mathbf{b})$ and $E(\mathbf{a'}, \mathbf{b'})$, which are

$$E(\mathbf{a}, \mathbf{b'}) = -cos(\theta) \qquad E(\mathbf{a'}, \mathbf{b}) = -cos(\theta) \qquad E(\mathbf{a'}, \mathbf{b'}) = -cos(3\theta) \quad (2.120)$$

By combining them together, we form quantum mechanical equivalent of Eq. 2.117, which we can now substitute into CHSH inequality defined in (2.118) as the average value $\langle S_{QM} \rangle$.

$$\langle S_{QM} \rangle = E(\mathbf{a}, \mathbf{b}) + E(\mathbf{a}, \mathbf{b'}) + E(\mathbf{a'}, \mathbf{b}) - E(\mathbf{a'}, \mathbf{b'}) \qquad (2.121)$$

This gives us a testable variant of Bell's inequalities. If we now set the angle according to figure Fig. 2.7

$$\theta_{ab} = \theta_{a'b} = \theta_{ab'} = \frac{\pi}{4} = 45° \qquad \theta_{a'b'} = \frac{3\pi}{4} = 135° \qquad (2.122)$$

we arrive at the result

$$\langle S_{QM} \rangle = -\cos\left(\frac{\pi}{4}\right) - \cos\left(\frac{\pi}{4}\right) - \cos\left(\frac{\pi}{4}\right) + \cos\left(\frac{3\pi}{4}\right) = -2\sqrt{2} \qquad (2.123)$$

Fig. 2.7 CHSH inequality
violation vectors with $\frac{\pi}{4}$
angle between each
neighboring pair of vectors

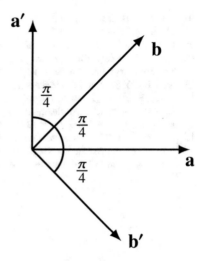

which is an obvious violation of CHSH inequality. This obtained value—a violation
of the inequality by a factor of $\sqrt{2}$—is called a *Tsirelson bound* and represents the
maximal violation of the CHSH inequality.

The conclusion to be drawn here is, yet again, that the statistical properties exhibited by independent measurements of entangled quantum states are stronger than
what could be pre-encoded in the particles upfront by any kind of hidden variable
parameter set. It is not possible to define a hidden variables scheme that will be able
to account for quantum correlations.

References

1. Bell, J. S. (1964). On the Einstein Podolsky Rosen paradox. *Physics Physique Fizika, 1*(Nov), 195–200.
2. Born, M. (1926). Zur Quantenmechanik der Stoßvorgänge. *Zeitschrift für Physik, 37*(12), 863–867.
3. Caves, C., & Fuchs, C. (1996). Quantum information: How much information in a state vector? 02.
4. Cerf, N. J., Adami, C., & Kwiat, P. G. (1998). Optical simulation of quantum logic. *Physical Review A, 57*(3), R1477–R1480.
5. Chen, M. -C., Wang, C., Liu, F. -M., Wang, J. -W., Ying, C., Shang, Z. -X., et al. (2022). *Ruling out real-valued standard formalism of quantum theory.*
6. Clauser, J. F., Horne, M. A., Shimony, A., & Holt, R. A. (1969). Proposed experiment to test local hidden-variable theories. *Physical Review Letters, 23*(Oct), 880–884.
7. Cleve, R., Ekert, A., Henderson, L., Macchiavello, C., & Mosca, M. (1999). On quantum algorithms.
8. Dirac, P. A. M. (1939). A new notation for quantum mechanics. *Mathematical Proceedings of the Cambridge Philosophical Society, 35*(3), 416–418.
9. Djordjevic, I. B. (2019). *Physical-layer security and quantum key distribution.* Springer Nature Switzerland.

10. Eberhard, P. H. (1977). Bell's theorem without hidden variables. *Il Nuovo Cimento B, 1971–1996*(38), 75–80.
11. Eddington, A. S. (1923). *The mathematical theory of relativity*.
12. Jaeger, G. (2009). *Entanglement, information, and the interpretation of quantum mechanics*. The Frontiers collectionBerlin: Springer.
13. Kok, P. (2018). *A first introduction to quantum physics*. Springer International Publishing.
14. Laloe, F. (2019). *Do we really understand quantum mechanics?* Cambridge University Press.
15. Li, Z.-D., Mao, Y.-L., Weilenmann, M., Tavakoli, A., Chen, H., Feng, L., et al. (2021). *Testing real quantum theory in an optical quantum network*.
16. Miller, J. L. (2022). *Does quantum mechanics need imaginary numbers?*.
17. Nielsen, M. A., & Chuang, I. (2010). *Quantum computation and quantum information: 10th anniversary edition*. Cambridge University Press.

Chapter 3
Getting Started with QDK and Q#

The content of this book up to this point provided the necessary quantum mechanical theoretical background. In particular, we covered historical development of the quantum theory, as well as provided an overview of its formulation and mathematics, thus establishing solid foundations upon which we can proceed further.

The final prerequisite needed before immersing ourselves in the theoretical and practical aspects of quantum computing is to go through the basics of Microsoft Quantum Development Kit, with the goal of establishing familiarity with the syntax and semantics of the Q# programming language. We shall focus mostly on the classical aspects of Q# and QDK at this point, and explicitly avoid diving too deeply into any quantum concepts in this chapter—we will have the rest of the book for that. The major objectives are ensuring successful setup of the development environment, as well as learning the core concepts of the Q# language. Readers who are already proficient with QDK and Q# may consider skipping this chapter—in any case, it can always be returned to and used as a reference.

We will reap the benefits of this chapter in the subsequent sections of this book, as it will enable us to freely switch between mathematical formalism and quantum program code written in Q#, equipping us with two complementary vectors for reasoning about quantum computing concepts and algorithms.

3.1 Quantum Development Kit

Quantum Development Kit from Microsoft, commonly referred to as the *QDK*, is a development toolchain around the Q# programming language, aimed at providing a holistic developer experience and state of the art environment for building applications that can be executed on quantum hardware. QDK consists of the Q# compiler, Q# language server and the Q# tooling in code editors, Q# standard library

© The Author(s), under exclusive license to Springer Nature Switzerland AG 2022

F. Wojcieszyn, *Introduction to Quantum Computing with Q# and QDK*, Quantum Science and Technology, https://doi.org/10.1007/978-3-030-99379-5_3

defining core features of the language, additional Q# libraries which can be used for extra research fields such as chemistry and machine learning, as well as quantum simulators, which can be used to test the programs on classical hardware.

The heart of the QDK is the Q# programming language, which is a statically typed, type safe high-level programming language from Microsoft Quantum. It abstracts away all of the complexities of interacting with the quantum hardware, surfacing only the familiar language constructs to the programmer.

Similarly to how cross-platform compilation can be done in other programming ecosystems, a program written in Q# is compiled against reference assemblies which, mostly, provide no implementation logic. Instead those intrinsic operations are later linked by the compiler to the target specific implementations. This way, a Q# program can be executed on a wide variety of quantum hardware platforms, and, conversely, not every operation exposed by standard Q# libraries may be implemented by every target platform.

At development time, QDK can be used to build quantum programs with Q# in three primary ways:

- as standalone applications
- using Jupyter Notebooks
- integrated into Python and .NET programs, using the interop between those languages and Q#

Each of these three approaches has its advantages and use cases, though for the purposes of this book, we will focus on *standalone Q# applications*, which are programs written entirely in Q# and executed from the command line interface with QDK tooling built around the .NET CLI. Regardless of that, all the Q# code discussed in this book is portable to any of the three methods mentioned above.

Execution from the command line is the natural, familiar path for many programmers, because it is consistent with the established patterns of working with many other programming languages. It can cater for programs of any size and complexity, and can be easily facilitated by the powerful Q# language editor extensions, which provide advanced code authoring experience.

Jupyter Notebooks is not suitable for complicated, large scale programs, but provide interactive environment which helps exploring the available APIs and play around with ideas, as well as interpolate Q# code with notes and other documentation. From that perspective, Jupyter Notebooks are very popular in academia, however due to the limitations in terms of IDE-like features, we will not be relying on them throughout this book.

Finally, language interop with Python and .NET allows building classical host programs, which focus on classical logic and can orchestrate various quantum algorithms. This tends to be the most rewarding path in real life scenarios, as it allows harnessing the full power of a mature high-level classical programming language, and supplement it with quantum logic written in Q#, but is also most complicated in terms of set up.

Table 3.1 Overview of the Q# project templates

Template name	Short name	Purpose
Console application	console	Generic Q# standalone app
Quantum application Honeywell	azq-honeywell	Standalone app for Honeywell QPU
Quantum application IonQ	azq-ionq	Standalone app for IonQ QPU
Class library	classlib	Reusable Q# library
xUnit Test Project	xunit	Q# unit test project, based on xUnit

3.1.1 Setting up the QDK

The prerequisite for getting started with the QDK is to install .NET SDK.[1] It is cross platform, free and open source package consisting of the .NET runtime and the .NET CLI published by Microsoft. The version required by the QDK at the time of writing is .NET SDK 6.0+, though this might change in the future. QDK is currently not compatible with ARM processor architecture, so even though .NET SDK for such platforms is available, an emulated x64 installation must be used.

Once the .NET SDK is installed, the quantum development templates need to be installed as well. This is done by executing the command shown in Listing 3.1. It will install the Q# project templates, which are summarized in Table 3.1.

```
dotnet new -i Microsoft.Quantum.ProjectTemplates
```

Listing 3.1 Installing the QDK templates into the .NET SDK

At this point a new Q# project, based on any of the templates, can be created in the current directory using the standard .NET CLI functionality to bootstrap new application as shown in Listing 3.2. A project can also be generated directly into a different directory using the -o command line option, and passing in the relative folder path.

```
dotnet new {template short name} -lang Q#

// example for a standalone Console Application
dotnet new console -lang Q#
```

Listing 3.2 Create a first Q# application using the .NET CLI

For the standalone console application the command generates a single Q# file `Program.qs` and a Q# project file whose name is inferred from the current folder. The code file is shown in Listing 3.3. We will cover the structure of Q# programs and semantics of the language in Sect. 3.2.

```
namespace first_quantum {

    open Microsoft.Quantum.Canon;
    open Microsoft.Quantum.Intrinsic;
```

---

[1] https://docs.microsoft.com/en-us/dotnet/core/sdk.

```
@EntryPoint()
operation HelloQ() : Unit {
    Message("Hello quantum world!");
}
}
```

Listing 3.3 Simplest possible Q# program as generated from the standalone console application template

Because Q# is so tightly integrated with the .NET development environment, Q# project files are effectively the standard .NET MSBuild files, and from that perspective look very much like C# project files (and even have `csproj` extensions). The reference to the specific QDK version is declared at the top of the generated project file, as seen in Listing 3.4.

```
<Project Sdk="Microsoft.Quantum.Sdk/0.21.2112180703">
  <PropertyGroup>
    <OutputType>Exe</OutputType>
    <TargetFramework>netcoreapp3.1</TargetFramework>
  </PropertyGroup>
</Project>
```

Listing 3.4 A basic Q# project file generated from the standalone console application template

The project file is used to manage and configure a magnitude of options, in particular declaring the dependencies of the Q# program. Those could be both first- and third-party libraries published to the official .NET package manager at nuget.org, or other Q# projects available locally on the development machine.

Both of them are added as two separate `ItemGroup` collection nodes under the `Project` node—external packages as `PackageReference` entries with the appropriate version from NuGet, while local projects as `ProjectReference` using a relative path. An example is shown in Listing 3.5.

```
<ItemGroup>
  <PackageReference Include="Microsoft.Quantum.MachineLearning"
    Version="0.21.2112180703" />
</ItemGroup>

<ItemGroup>
  <ProjectReference Include=
    "..\first-quantum-library\first-quantum-library.csproj" />
</ItemGroup>
```

Listing 3.5 Referencing Q# dependencies

The core functionalities provided by Q# are included in the QDK by default, as part of an implicit `Microsoft.Quantum.Standard` package reference. This reference is not visible in the generated project file, but is resolved by the QDK itself. From that perspective, all of the core features of the ecosystem covered in this book are available in any Q# project out-of-the-box, and no additional dependency declarations are required.

Aside from the standard Q# library Microsoft publishes three additional official libraries that can be used for Q# development—`Microsoft.Quantum.`

`Chemistry` for quantum chemistry, `Microsoft.Quantum.Machine Learning` for quantum machine learning and `Microsoft.Quantum. Numerics` for quantum arithmetic.

Interestingly, the **QDK itself never needs to be installed explicitly**. Instead, such newly bootstrapped quantum program can be run locally using the regular `dotnet run` command, used to compile and run .NET applications. This command implicitly restores all the dependencies of the executed program, including installing the QDK. It then builds the application using the Q# compiler, and finally invokes it on the default simulator shipping with the QDK. For the default Q# standalone console application, it should produce:

```
> dotnet run
Hello quantum world!
```
Listing 3.6 Running the default Q# project from the command line.

The output to the standard I/O is facilitated by using a core Q# operation `Message` (`msg : String`) from the `Microsoft.Quantum.Intrinsic` namespace, which allows logging any string-based output from within the program. Its behavior can vary from simulator to simulator, as the target to which the message would be written may be different. We shall use this operation a lot throughout the book, because it provides the easiest way to output or visualize data from a quantum application.

This should validate the successful set up of the QDK. When running the programs locally on the development machine this way, the QDK uses one of its simulator environments as the execution target. Those are covered in Sect. 3.5.

The project file can also be used to instruct the compiler for which target quantum hardware the program needs to be compiled for. By default, the entire API surface of the Q# standard library and any valid Q# package reference are allowed to be freely used. At the same time, current generation of quantum hardware still supports only a subset of the available instruction set—in fact, at the moment no vendor provides a quantum computer capable of running a Q# program taking full advantage of all Q# language constructs and built-in APIs.

While Microsoft does not offer first-party quantum hardware to the public yet, Q# programs can be specifically compiled for IonQ and Honeywell quantum processing units by adding the `ExecutionTarget` property to the Q# project file, as shown in Listing 3.7. At the time of writing, the relevant values that can be set there are `ionq.qpu` and `ionq.simulator`, as well as `honeywell.qpu` and `honeywell.simulator`, and are dictated by the platforms available in Azure Quantum. As more hardware vendors embrace Q# (more on that in Sect. 3.3), more execution targets will become available.

```
<Project Sdk="Microsoft.Quantum.Sdk/0.21.2112180703">
  <PropertyGroup>
    <OutputType>Exe</OutputType>
    <TargetFramework>netcoreapp3.1</TargetFramework>
    <ExecutionTarget>ionq.qpu</ExecutionTarget>
  </PropertyGroup>
</Project>
```
Listing 3.7 Q# project file configured for running on the IonQ hardware.

If a certain instruction set is unavailable for the given execution target the program will not compile and the compiler will issue the appropriate error explaining the problem, such as for example a hint that conditional expressions cannot be executed on the selected quantum hardware. For practical purposes we shall not dive into the limitations and restrictions of specific quantum computing platforms in this book, because it is a very volatile area, with rapid progress and constant changes in the hardware. It is not unreasonable to expect that in the future most of these restrictions will disappear or become significantly less constraining.

3.1.2 Setting up the IDE

It is one thing to bootstrap the command line environment for generating, compiling and running Q# programs, but it is a separate problem to configure the code editor environment for convenient development activities.

Microsoft Quantum team publishes two official first-party IDE extensions, one for Microsoft Visual Studio[2] and one for Visual Studio Code. In particular Visual Studio Code,[3] which is cross platform, fully open sourced and MIT licensed, has become extremely popular among software developers over the past several years. In fact, despite being relatively new, compared to other popular editors with storied histories like Vim, Emacs or Notepad++, it is already the preferred editor choice for the majority of programmers [8].

While for the purposes of this book just about any editor would work—after all, the Q# compiler is accessible from the command line—it is strongly recommended to proceed forward with Visual Studio Code as the editor experience. Once Visual Studio Code is installed, the Q# extension can be installed from the extension marketplace, where it bears the official name **Microsoft Quantum Development Kit for Visual Studio Code**.

Interestingly, QDK also provides a standalone Q# language server implementation compatible with the specification of the Language Server Protocol (LSP) [5]. Almost every modern editor or IDE has the capability of easily interacting with LSP, and therefore, at least in theory, it should not be that much work to wire in Q# support as a custom extension or plugin of the reader's favorite editor of choice.

The Q# language experience in the editor is very rich—it uses the compiler at design time and is therefore able to understand the semantics of the program being authored. It provides powerful code completion, code navigation, real-time diagnostics and automated code actions. Additionally, a compiler-based code formatter is part of the extension as well. The quality of tooling may not be surprising to

---

[2] Microsoft Visual Studio is available for Windows only. Confusingly, the Visual Studio available for MacOS and Linux is marketed as the Unix version of Windows product, but from the technical perspective is a completely different IDE. It is not supported by the QDK.

[3] https://code.visualstudio.com.

developers with .NET background, because in general the language services and developer experience for all .NET languages are excellent, but compared to some other programming languages, Q# editor experience is really exquisite.

3.2 Q#

One of the major advantages of using a high-level programming language like Q# for quantum programming is that it facilitates hybrid classical-quantum program composition. Q# enables quantum software authors to think and structure their programs much like classical programing is done.

In Listing 3.3, we introduced the most basic possible Q# program, printing a simple Hello world! message out to the console. In fact, that program does not take advantage of any quantum programming features and could be easily executed on a classical machine only, it does however already hint at the general syntactic flavor of the language. We will now go through the most important features and design paradigms of the language. For reference, the latest complete Q# language specification is always available in the official GitHub repository of the language [6].

3.2.1 Core Language Concepts

3.2.1.1 Declarations and Scope

Variables in Q# are declared using the let and mutable keywords. The former creates an immutable value binding which prevents a variable from being reassigned, while the latter allows changing the value of the given variable within the lifetime scope of the variable. Reassigning a variable value is done using a dedicated set keyword in what is referred to as a *copy-and-update expression*. The variable type is always inferred and, generally speaking, no type annotation is required.

An example of declaring immutable and mutable variables is shown in Listing 3.8.

```
// declare text1 as immutable
let text1 = "hello";

// declare text2 as mutable
mutable text2 = "hello";

// change its value
set text2 = "world";
```

Listing 3.8 Mutable and immutable local declarations in Q#.

The scope for the declarations is always defined by the surrounding statement block, delimited by curly braces.

3.2.1.2 Conditionals

Q# supports only `if`-based conditional branching. An if-statement evaluates the supplied boolean logical condition and creates a locally scoped block in which desired statements can be placed. Variables declared within this block are not visible outside of it. An if-statement can be supplemented with alternative execution paths using the `elif` blocks, which can evaluate any number of logical conditions, and an `else` block, which is the catch-all fallback for all the other cases not matched by the supplied logical expressions.

In terms of constructing those logical boolean expressions, Q# supports `and`, `or` and `not` keywords. Equality can be verified using the `==` and `!=` operators. Comparative expressions are limited to types that support equality—numeric types, strings, booleans, as well as the quantum-specific data types. In particular, tuples and arrays do not currently support equality comparison.

Quantitative comparison between various numeric types can be done using the usual set of greater-than >, greater-than-or-equal >=, less-than < and less-than-or-equal <= operators.

If-statements do not require the logical conditional to be wrapped in parentheses. An example of an if-statement with the alternative path blocks is shown in Listing 3.9.

```
if number > 10 {
    // logic to execute when
    // number is bigger than 10
} elif number > 0 {
    // logic to execute when
    // number is bigger than 0 and smaller or equal to 10
} else {
    // logic to execute when
    // number is smaller or equal to 0
}
```
Listing 3.9 A conditional block.

Logical blocks consisting of if-else branches only, can be simplified using the conditional expression, such as the one shown in Listing 3.10. The operator used there is often referred as the ternary operator and consists of the boolean condition, followed by a question mark ?, which denotes the expression to be evaluated when the condition is met, and further followed by a pipe symbol |, which indicates the expression to be evaluated when the condition is not satisfied.

```
let text = number > 10 ? "number > 10" | "number <= 10";
```
Listing 3.10 A conditional expression.

3.2.1.3 Loops

The primary way of performing iterations in Q# is by using `for` loops. They can be executed against a range of values, which corresponds to the typical behavior in other programming languages, or they can be set up to directly iterate through items of an array. The latter behavior is often expressed as *foreach* in other languages, though Q# does not require such explicit keyword distinction.

Q# loops are subject to the same scoping rules we already covered, and similarly to conditional statements, the syntax does not require extra parentheses. Q# does not support early `for` loop termination—they always need to fully complete, and cannot be broken out of. With regards to that, Q# supports `while` loops, but they can only be used inside of functions, and not in quantum operations.

An example of iterating over a range, where the index is used to access array elements, as well as iterating directly over an array, is shown in Listing 3.11. These approaches are semantically equivalent and a matter of preference.

```
let numbers = [1, 2, 3, 4, 5];
for index in 0 .. Length(numbers)-1 {
    // can access array item via numbers[index]
}

for item in numbers {
    // can access array item via item variable
}
```

Listing 3.11 Two variants of for loops in Q#.

In addition to the `for` loops, Q# also has the concept of conditional loops, expressed via the `repeat-until-fixup` statements as shown in Listing 3.12. They allow to repeat a specific instruction sequence until a certain condition, classical or quantum (measurement-related), is fulfilled. The logic to be repeated is captured in the `repeat` block, while the termination condition is evaluated as a single boolean `until` statement. The `fixup` block is optional and executes prior to each subsequent iteration.

```
mutable counter = 1;
repeat {
    // logic to execute on each iteration
    set counter += 1;
} until counter > 5
fixup {
    // logic to execute after each run
    // except for the final one
}
```

Listing 3.12 A repeat-until-fixup loop in Q# with five iterations.

Conditional loops are very complicated to be implemented on today's quantum hardware, therefore even though they are supported by Q# as part of the language specification, they currently cannot be used when compiling for IonQ and Honeywell hardware.

3.2.1.4 Namespaces

Namespaces are the only block structures allowed at the top level of the program, and are a prerequisite for defining any functions, operations or types, as all of those must be contained within a namespace. Therefore a namespace declaration statement is normally the first thing that is defined in a Q# program and that is encountered in each Q# file.

A program usually consists of multiple namespaces, though a single shared namespace is also possible, especially as namespaces are allowed to span multiple files. They do provide a logical grouping of declarations that they contain, as everything within a given namespace is automatically visible to all the other code paths found within that same namespace.

To access a symbol from an external namespace, a *fully qualified* name must be used, consisting of the symbol name, preceded by its namespace—for example `Microsoft.Quantum.Convert.{symbol name}`. However, this is rather verbose and hardly practical, so, alternatively, a namespace can contain references to other namespaces via the open directive. In that case, all symbols contained in the opened namespace also become visible and are accessible without an explicit need to fully-qualify them. An example is shown in Listing 3.13.

```
namespace Filip.Quantum.Sample {

    // all symbols from this namespace are available as if
    // they were declared in Filip.Quantum.Sample - via
    // the unqualified names
    open Microsoft.Quantum.Convert;

    // declarations of types, operations, functions go here
}
```
Listing 3.13 A Q# namespace.

When opening a lot of namespaces, it is possible to run into naming conflicts between different symbols. It may also be overwhelming or confusing to have a large amount of symbols visible directly that way. To remedy both of these issues, it is possible to alias a namespace when opening it using the `as ...` syntax construct. Once a namespace is opened with an alias, the alias becomes a simplified qualification for the symbol, and must prefix any interaction with a symbol from that namespace, as shown in Listing 3.14.

```
namespace Filip.Quantum.Sample {

    // using Convert.{symbol name} simplified qualification
    open Microsoft.Quantum.Convert as Convert;
}
```
Listing 3.14 Aliasing an opened Q# namespace.

3.2.1.5 Conjugations

Conjugation support in Q# is a quantum-specific language feature. Given a transformation $\mathbf{V}$ which is preceded by a transformation $\mathbf{U}$ and succeeded by a transformation $\mathbf{U}^\dagger$, we can use the Q# conjugation statement for our convenience.

$$\mathbf{U}^\dagger \mathbf{V} \mathbf{U} \tag{3.1}$$

In this case, we perform a specific computation $\mathbf{U}$, to temporarily put a system in a certain quantum state, on which we would like to execute transformation $\mathbf{V}$. Afterwards, by invoking $\mathbf{U}^\dagger$ we effectively uncompute the initial unitary transformation. A common example of this programming pattern is computation in a different basis than the Z-basis—as conjugation language feature would facilitate temporary basis switch, application of certain set of unitary transformations in the new basis, and then basis change back to standard basis.

Syntactically, conjugations take the form shown in Listing 3.15, with the `within` block acting as $\mathbf{U}$ and $\mathbf{U}^\dagger$, and the `apply` block corresponding to the $\mathbf{V}$. However, since conjugations are an exclusively quantum feature, we shall not explore them any further in this chapter, and return to them in Chap. 4.

```
within {
    // prepare state / unprepare state
}
apply {
    // compute
}
```
Listing 3.15 Q# conjugation statement.

3.2.2 Types

Compared to general purpose programming languages, Q# is characterized by a rather basic type system. It has a strongly functional flavor to it, and therefore feels much more natural to software engineers with background in that class of languages. Q# is nominally typed, meaning that types are considered equivalent based on their name, rather than based on their internal structure. In addition to that, Q# uses a very powerful type inference mechanism and, wherever possible, automatically determines the type of a given declaration.

All the types in the language are value types and Q# explicitly does not have the concept of pointers or references. Because of that, it is not possible to encounter the dreaded null reference exception in Q#—*nulls* simply do not exist. It also does not have a concept of functions returning *void*, instead, a no-op `Unit` type is used to decorate functions that do not return a usable value.

Table 3.2 Overview of the built-in Q# types

Type	Characteristics	Default value
String	Encoded in UTF-16	""
Int	64-bit, signed	0
BigInt	Any bit length, signed	0L
Double	64-bit floating point numbers	0.
Bool	Single bit	false
Unit	Void type	()
Qubit	Quantum bit	invalid state
Result	Quantum measurement result	Zero
Pauli	Pauli observable	PauliI
Range	Sequence of integers	[]
Array	Typed set of values	[]
Tuples	Grouping of values of any type	default of each element

3.2.2.1 Basic Types

Q# has only a handful of built-in types. There are three numerical types—Int, BigInt and Double, allowing developers to express integers of different bit length and floating point numbers. String values are captured by the String type, while logical statements are used with expressions evaluating to Bool type.

Table 3.2 shows the summary of all of the core built-in types available in Q#.

In addition to that, as we will learn soon, there are several types supplied by the core library of Q# that fall under the so-called *user-defined types* category.

3.2.2.2 Quantum Domain

The Q# language has three main built-in types related to the quantum domain. The first one is the Qubit, which is the most basic unit of information in quantum information theory and thus the most fundamental way of interacting with the quantum state of the program. Q# does not require the program author to understand whether the underlying qubit at the hardware level is available as logical or physical—this is abstracted away and is considered to be the explicit responsibility of the runtime.

When authoring a quantum program, the most fundamental operation one might perform is to allocate a qubit—it is, after all, the prerequisite to performing any further computations. Q# has two separate concepts when working with qubits— allocation through the use keyword and the borrow keyword. They can be followed by an optional statement block, which controls the lifetime of the qubit as part of the scope associated with that block. If that block is omitted, the qubit is available until the end of the closest surrounding scope.

Allocating a qubit with the use statement means allocating a fresh qubit from the logical pool that is available on the quantum hardware. The provisioning mechanism is deterministic—such qubit is guaranteed to be in state |0⟩ and any quantum algorithm can safely rely on that. On the other hand, a qubit allocated through the borrow statement does not give the developer any promises about the state of the qubit, it will be in a *dirty* state. Allocating qubits through borrowing is less common for beginner exploration, but is very useful in certain algorithms where extra auxillary qubits might be needed, for example to ensure reversibility of a given operation. Borrowing, instead of fully allocating qubits can considerably reduce quantum memory footprint of the program.

As dictated by the laws of quantum mechanics, the qubit state is opaque and not possible to be introspected in any way at runtime—the only way for the program to extract classical information out of it is to apply an observable, using the various measurement operations exposed by the Q# standard library. This limitation applies to running on quantum hardware—on the other hand, for programs executed on quantum simulators, Q# provides debugging tooling that allows dumping the quantum state of a program and performing full analysis of the state of the involved qubits. We shall return to those concepts in Chap. 4.

The other two quantum-specific built-in types in Q# are Result and Pauli, both of which are related to the process of measurement of the qubit. Pauli represents the four Pauli observables, and can be used via the built-in keyword literals PauliX, PauliY, PauliZ and PauliI. Result, on the other hand, represents the measurement result using a given quantum operator, with eigenvalues of $+1$ or -1. The possible values are Zero and One, and they correspond to the convention used in quantum computing. They are also both available as keyword literals and can therefore be used in a strongly typed fashion.

We mention these quantum-specific concepts only briefly at this point, as we shall cover all of them in detail in Chap. 4. Naturally, beyond those basic types, Q# standard library and related QDK packages also come with an abundance of built-in algorithms, features and operations specifically included to make implementation of quantum workflows easier. We shall explore and use many of them as we progress in this book.

3.2.2.3 Collections

Contrary to many other programming languages with a wide spectrum of specialized collection types, Q# only has two—Range and Array.

Range is a sequence of integer values which is suitable for iteration—it is therefore often used together with for loops. Ranges can be created using the range expression which has the format of start..step..end. Step determines the increment size between subsequent range elements. It is optional and can be omitted—the compiler defaults it to 1. If the end is larger than start, and empty range is created, unless the step size is defined as negative, in which case the generated sequence of integers would be decreasing. Both start and end values are included into the range. Some simple examples of ranges are shown in Listing 3.16.

```
let range1 = 1..4; // 1, 2, 3, 4
let range2 = 1..2..5; // 1, 3, 5
let range3 = 3..-1..1; // 3, 2, 1
let range4 = 3..2; // empty
```
Listing 3.16 Range examples in Q#.

Arrays can be used to store instances of the same type—both classical ones, such as string or integers, as well as quantum ones, most typically an array of qubits.

Out of the box Q# provides a very rich set of array manipulation functions, under the `Microsoft.Quantum.Arrays` namespace. The namespace contains ready-to-use functions to filter, fold, map, sort, transpose, search, zip an array, and many more. An array is created using the array creation expression which takes the form shown in Listing 3.17.

```
let labels = ["green", "red", "blue"];
```
Listing 3.17 Basic array creation in Q#.

An empty array can be created without any type annotation using initialization syntax [], where the type will be inferred by the compiler upon first usage of the array. If such array is not used anywhere in the program the compiler will not be able to infer its type and the program would not compile. Both arrays and ranges have default values of an empty set [].

Arrays respect the same immutability rules as the rest of the language and therefore, once created using the immutable `let` binding, an array as a whole, as well as its individual elements cannot be modified. A `mutable` binding, on the other hand, allows changes to the array using copy-and-update expression, creating a de-facto new array as a result. An example is shown in Listing 3.18.

```
// ["", "", ""]
mutable labels = ["", size = 3];

// ["", "blue", ""]
set labels w/= 1 <- "blue";
```
Listing 3.18 Copy and update of an array element in Q#.

Very conveniently, arrays can be easily concatenated together using the addition + operator.

```
let labels1 = ["green", "red"];
let labels2 = ["blue", "yellow"];

// ["green", "red", "blue", "yellow"]
let labels3 = labels1 + labels2;
```
Listing 3.19 Array concatenation in Q#.

An interesting application scenario for ranges is that they can be used to slice arrays. Given an array, the range can be used to determine which indices of the array to copy into a new array.

```
let labels = ["green", "red", "blue", "yellow"];
```

```
// ["red", "blue"]
let slicedLabels1 = labels[1..2];

// ["green", "blue"]
let slicedLabels2 = labels[0..2..4];

// ["blue", "red", "green"]
let slicedLabels3 = labels[2..-1..0];
```

Listing 3.20 Array slicing using a range.

While Q# does not allow open-ended ranges, and start and end of a range must normally be specified in the range expression, this requirement is relaxed when slicing an array using a range. In such situations, if the range does not have a start or end, those get inferred automatically from the size of the array being sliced. For open-ended ranges the start or end values are replaced with a placeholder dot . symbol.

```
let labels = ["green", "red", "blue", "yellow"];

// ["green", "red", "blue"]
let slicedLabels1 = labels[...2];

// ["red", "yellow"]
let slicedLabels2 = labels[1..2...];

// ["yellow", "blue", "red", "green"]
let slicedLabels3  = labels[...-1...];
```

Listing 3.21 Array slicing using an open-ended range.

3.2.2.4 Tuples

Because the entire Q# type system is quite basic, an important role in it is played by tuples, as they allow grouping items of various types into composite structures. Tuples can be used as locals and are permitted in callable signature definitions— as both arguments and return values. Tuples can be fully or partially deconstructed into local variables using a local declaration statement. It is possible to omit certain values upon deconstruction using the discard _ symbol. An example of a function returning a tuple of two elements is shown in Listing 3.22. The function relies on the built-in conversion methods from the Microsoft.Quantum.Convert namespace to perform type casting.

```
function Tuples() : Unit {
    let result = TupleFunction(100, true);
    let (num1, txt) = TupleFunction(200, true);

    // with discard - deconstruct only part of the result
    let (num2, _) = TupleFunction(300, false);
}
```

```
function TupleFunction(number : Int, flag : Bool) : (Double,
    String) {
    let result = (IntAsDouble(number), BoolAsString(flag));
    return result;
}
```
Listing 3.22 Tuples in Q#.

In Q#, a single element tuple is considered to be equivalent to a standalone item of the same type, a concept referred to as *singleton tuple equivalence*. This is a very powerful feature, because it allows developers to either omit or inject the tuple construct in various situations. For example, a callable with three arguments is equivalent to a callable with a single tuple argument, consisting of the same three elements, and can be invoked both ways. Such language characteristic has another useful consequence—it allows to think of all callables in purely functional fashion. They can all be reasoned about as analogous to mathematical function, always accepting a single parameter, and always returning a single parameter—it's just that parameter can be a tuple of multiple elements.

Listing 3.23 shows how we can call the same function as declared in Listing 3.22 by taking advantage of singleton tuple equivalence, and passing the method arguments using a tuple instead of individual parameters.

```
let items = (100, true);
let (num, txt) = TupleFunction(items);
```
Listing 3.23 Singleton tuple equivalence in Q#.

3.2.2.5 User-Defined Types

In addition to the built-in types, Q# also allows creation of the so-called *user-defined types*. The name is somewhat confusing, because it implies that they are only declared by the program's author in the application code. While this is often the case indeed, there are also types, commonly used in various algorithms, that are shipping together with the core libraries of Q#, such as `LittleEndian` or `DiscreteOracle`, which are also user-defined types.

User-defined types are not fully-fledged types as known from object-oriented programming languages, but instead are de-facto pure data types, without any behavior attached to them. This is similar to the concepts of structs or record types from other programming languages, which are simply grouping certain values into a set of fixed fields.

User-defined types are declared without constructors—instead there is always a single constructor available, one that requires all the type members to be supplied, and it is auto-generated by the compiler. A user-defined type is declared using a `newtype` keyword. An example is shown in Listing 3.24.

```
newtype LabeledResult = (
    Label : String,
```

```
MeasurementResult : Result);
```
Listing 3.24 Sample Q# user-defined type.

Properties of such a user-defined type can then be accessed using the item access operator : : or through the deconstruction operator !. Deconstruction allows unwrapping the entire user-defined type into a dedicated tuple. Both of these approaches are shown in Listing 3.25.

```
let result = LabeledResult("measurement one", Zero);

// either
let label = result::Label;
let measurementResult = result::MeasurementResult;

// or
let (label, measurementResult) = result!;
```
Listing 3.25 Accessing members in user-defined type.

User-defined types in Q# are immutable, and behave as value types instead of pointers. There is no way to update an instance of an already instantiated type—instead a new instance must be created.

3.2.2.6   Type Casting

While Q# allows conversion between certain types, it does not have an explicit cast operator, nor does it support implicit casting between different types. Because of that, it is not possible to use, for example, a double instead of an integer as part of the range definition. Instead, type conversion is possible via a predefined set of helper functions defined as part of `Microsoft.Quantum.Convert` namespace, allowing the developer to convert between the most common pairs of types, such as integers and doubles, ranges and arrays or booleans and strings, to name a few. There is no built-in support for conversion between user-defined types—those need to be coded manually should the need arise.

A simple example with a conversion of double to an integer is shown in 3.26.

```
let dblNumber = 1.;
let intNumber = DoubleAsInt(dblNumber);
```
Listing 3.26 Q# basic type conversion.

A special case for type conversion is the so called *string interpolation*. Q# allows using arbitrary expressions inside the specially annotated strings, such as that shown in Listing 3.27, without an explicit call to any conversion function. These expressions are evaluated when the string is rendered, and if the resulting type is convertible into a string directly, that conversion will be executed, otherwise the typical string representation for a given type is used. Interpolated strings look just like regular strings, except they are prefixed with $, while the individual embedded expressions to be evaluated are delimited with curly braces {}—in that sense, the implementation of string interpolation in Q# is identical to that in C#.

```
let index = 1;
let result = One;
let data = [1, 2, 3];

// outputs "Result of trial 1 is One. Data: [1,2,3]."
Message($"Result of trial {index} is {result}. Data: {data}.");
```
Listing 3.27 Q# string interpolation.

String interpolation is the only situation in which Q# would perform a de facto implicit conversion.

3.2.3 Callables

As it is the case in functional programming languages, behavioral logic in a Q# program is entirely contained in the so-called *callables*, which are analogous to functions in classical programming. Callables can be defined as Q# *functions* or Q# *operations*. They share most of the syntactic and semantic characteristics, though, as we shall see in a moment, there are some fundamental differences between the two as well.

All callables consist of a signature, defining the arguments that can be passed in, as well as the return type of a callable, and a body determining the local scope for declarations. Callables are required to have an execution exit point, which can be implicit or explicit. For callables returning a Unit value, which is equivalent to void in many other programming languages, the exit point is implicitly compiler-generated at the end of the callable scope as return (); and can be omitted by the programmer. All other return types require an explicit return statement. For complex logical paths, multiple return statements per callable are supported.

Arguments passed into the callables are always bound immutably and therefore behave exactly the same as local variables declared with the let keyword. Their values cannot be reassigned and their visibility and lifetime is bound to the scope of the callable.

3.2.3.1 Functions

Functions are used to express classical deterministic logic and are evaluated as soon as they are invoked. Because of that characteristic functions are not allowed to interact with the quantum state of the program. An example of a function verifying if XOR between a pair of boolean arguments is equal to the OR between them is shown in Listing 3.28.

```
function XorEqualsOr(x : Bool, y : Bool) : Bool {
    return Xor(x, y) == (x or y);
}
```
Listing 3.28 Sample Q# function.

Functions do not have side-effects and always produce the same output for the same input. As such, no matter how many times we invoke the function, if the same parameters are passed in, the return value would be unchanged. Functions can also call other functions, which is what the example does, as it makes use of the built-in Q# function Xor (a : Bool, b : Bool) from the Microsoft.Quantum.Logical namespace.

3.2.3.2 Operations

Operations are used to interact with the quantum state of the program, through native instructions supported by the given targeted quantum processor. As a result they must have side-effects, since a quantum state transformation causes the change of the global state of the program which is external to the actual operation.

A useful way of thinking about functions and operations, is that functions capture deterministic classical logical, while operations represent quantum logic to be executed against the quantum program state. A more generalized perspective is to view functions as a way to *compute*, while consider operations as a mechanism to *do* something [4].

A sample Q# operation interacting with a quantum state is shown in Listing 3.29. We shall not dive into the details of it yet, because we will dedicate much of the rest of the book to it, but in this case the operation allocates and measures a qubit.

```
operation AllocateAndMeasure() : Result {
    use qubit = Qubit();
    return M(qubit);
}
```
Listing 3.29 Sample Q# operation.

Because of their deterministic nature, functions cannot call any operations, but the opposite is allowed—operations can always invoke any functions they need to. Callables can be declared in any order—contrary to some other programming languages, it is not necessary to explicitly declare a callable before invoking it—as long as it would still be part of the same compilation unit.

It is not possible to throw exceptions from within a callable—Q# does not have any exception types nor does it provide any try/catch infrastructure. An alternative, however, is to use a fail statement to abort the execution of the program completely. A fail statement is treated by the compiler as a valid replacement for a return statement, namely a callable containing a fail statement would compile successfully even when it is missing a mandatory return statement.

```
function Fail() : Unit {
    fail "Something bad happened!";
    Message("Hello"); // this line won't be reached
}
```
Listing 3.30 Q# fail statement.

3.2.3.3   Attributes

Operations and functions can be annotated with attributes providing extra metadata for the compiler, the runtime or for design-time tooling. A few of them are built-into the Q# standard library, and located under the `Microsoft.Quantum.Core` namespace. An attribute is placed directly above the callable signature and preceded by the @ character and followed by () parentheses.

Attributes provide no behavior on their own, but their presence can be freely interpreted by the Q# ecosystem and tooling. For example, the `EntryPoint` attribute is used to instruct the compiler to start the Q# program execution specifically from this marked operation. Once discovered, the compiler emits the necessary instructions to facilitate that. By convention, only one such entry point operation per program is allowed, and using this attribute more than once will result in a compilation error.

```
@EntryPoint()
operation Main() : Unit {
    // program logic
}
```
Listing 3.31   Entry point operation Q#.

Attributes can also be created by the user using user-defined types. In fact, any user-defined type becomes an attribute once it is annotated with `Microsoft.Quantum.Core.Attribute` attribute.

```
@Attribute()
newtype MyAttribute = (Text : String);

@MyAttribute("Foo")
@EntryPoint()
operation Main() : Unit {
    // program logic
}
```
Listing 3.32   Declaring a custom attribute.

Multiple attribute annotation on the same callable are allowed.

3.2.3.4   Operation Specializations

Two common unitary transformation patterns used in quantum information theory are adjoints and controlled transformations. Adjoint of an arbitrary transformation U is its conjugate transpose $U^{\dagger}$.

Since a Q# operation is a quantum state transformation and from the theoretical perspective, if it does not perform a measurement, can be treated as such arbitrary $\mathbf{U}$, we can instruct the Q# compiler, by adding an *operation characteristic* Adj to the operation signature, that it could automatically generate its adjoint $\mathbf{U}^\dagger$. This allows this operation to be invoked as regular $\mathbf{U}$ or as its inverse $\mathbf{U}^\dagger$.

The controlled execution, on the other hand, is a concept used in various quantum algorithms, and relates to conditional application of $\mathbf{U}$. Similarly, to an adjoint, we can ask the compiler to support us to automatically generate the controlled specialization. This, in turn, enables us to invoke it standalone or together with the control condition.

Operation characteristics follow the operation signature and are declared using the is keyword. Adjoint specialization is represented by the Adj characteristic, while controlled specialization is declared with Ctl characteristic. The example of declaring both specialization is shown in Listing 3.33.

```
operation QuantumTransformation(q : Qubit) : Unit is Adj + Ctl {
    // actual transformation logic
}
```

Listing 3.33 Q# operation with both adjoint and controlled operation characteristics.

Operation characteristics are the most common and convenient way of declaring specializations, though not the only one. It is also possible to instruct the compiler to create them from within the operation block. Such inline approach, semantically equivalent to that from Listing 3.33, is shown in Listing 3.34. Note the extra separate block needed to host the body of the operation now.

```
operation QuantumTransformation(q : Qubit) : Unit {
    body (...) {
        // actual transformation logic
    }
    adjoint auto;
    controlled auto;
    controlled adjoint auto;
}
```

Listing 3.34 Q# operation with inline adjoint and controlled specializations.

Finally, in addition to the compiler-generated adjoint and controlled specializations, for more advanced used cases they can also be manually declared, using dedicated adjoint and controlled blocks. The presence of the three dots ... in these declaration are needed to tell the compiler that the same arguments that are passed to the operation, should be passed to the specialization as well. Controlled specialization receives an extra argument corresponding to the control condition too.

```
operation QuantumTransformation(q : Qubit) : Unit {
    body (...) {
        // actual transformation logic
    }
    adjoint (...) {
        // adjoint logic
    }
    controlled (control, ...) {
```

```
        // controlled logic
    }
}
```

Listing 3.35 Q# operation with manually defined adjoint and controlled specializations

We shall rely on these specialization a lot throughout the upcoming parts of this book.

Once the specializations are declared, they can be accessed using *functors*. Q# supports two of them, `Adjoint` and `Controlled`, corresponding, naturally, to the two specialization types. A functor is a functional programming concept, with roots in the category theory, which performs mapping between two categories. In Q#, the functor is used to access the operation's specialization. Both of the Q# functors commute so the order of their application does not matter.

Listing 3.36 shows execution of both specializations using the respective functors, based on the operation defined in Listing 3.33. The controlled functor additionally requires an extra control qubit passed in separately (in the form of an array, since multiple qubits can act as controls too) as first argument.

```
Adjoint QuantumTransformation(qubit);
Controlled QuantumTransformation([controlQubit], qubit);
```

Listing 3.36 Invoking the specializations of an operation.

3.2.3.5 Callables as Arguments

As one would expect from a functional programming language, callables can also be passed as arguments to other callables. The syntax used for that is <TInput> => <TOutput> for an operation reference and <TInput> -> <TOutput> for a function reference.

An example is shown in 3.37, where a Q# standard library `FactorialI` (n : Int) function is wrapped with a custom function allowing the execution of an arbitrary preparation logic prior to calculating the factorial. This preparation logic is in this case passed in as a delegate argument.

```
function FactorialWithPreparation(
    input : Int,
    preparation : (Int -> Int)) : Int {
    let prepared = preparation(input);
    return FactorialI(prepared);
}

function Square(number : Int) : Int {
    return number * number;
}
```

Listing 3.37 Passing Q# function as function argument.

It can then be invoked by calling

```
let result = FactorialWithPreparation(2, Square);
```

Listing 3.38 Combining the two functions from 3.7 in a statement.

3.2.3.6 Partial Application

One particularly useful feature of Q# implementation of functions and operations is that they both support the concept of *partial application* [2]. Partial functions are integral aspect of all functional programming languages and allow dynamic transformation of a given method, without its evaluation, into a new reduced one, which accepts fewer parameters than the original. In other words, given a function or operation whose signature requires n arguments to be passed, the developer can supply $m < n$ of them, and obtain a new function definition as a result—one that requires $n - m$ parameters, and which can be invoked at a later stage.

Consider the built-in Q# function, from the Microsoft.Quantum.Arrays namespace, which allows padding an array to a desired size with a predetermined default value. Its signature is shown in 3.39.

```
function Padded<'T> (nElementsTotal : Int, defaultElement : 'T,
    inputArray : 'T[]) : 'T[]
```

Listing 3.39 The signature of one of the Q# standard library functions which could be used with partial application.

Because the function defines three parameters, we could partially apply it by supplying one or two of them upfront, and obtain a new function this way. The new function is then invoked by passing in only the missing parameters. This process is shown in Listing 3.40.

```
let values = [1, 2, 3, 4, 5];
let padWithZero = Padded(_, 0, _);

// [0,0,0,0,0,1,2,3,4,5]
let padded = padWithZero(10, values);
```

Listing 3.40 Partial application of the function from Listing 3.39.

Singleton tuple equivalence principle applies to partial callable application as well. Once such a function gets reduced to a single tuple in its signature, it can be called by omitting the tuple structure. This is shown next.

```
function Tuples() : Unit {
    let reducedFunction = ComplexTupleFunction(5, _);

    // works - but tuple can be omitted
    reducedFunction(("foo", true));

    // cleaner approach thanks to singleton tuple equivalence
    reducedFunction("foo", true);
}
```

```
function ComplexTupleFunction(number : Int, (data : String, flag
    : Bool)) : Unit {
}
```
Listing 3.41 Partial application of with singleton tuple equivalence.

Declaring callables as anonymous functions, or the so-called lambdas—inline variable function literals, is currently not supported but is in active development and will be added to Q# in 2022.

3.2.4 Generic Programming

Q# has support for generic typing for both operations and functions. From the syntactic point of view, type parameters are supplied in the callable signature and need to be prefixed with a single-quote character '. They are allowed both in the input as well as in the output parameters of the callable.

Many of the built-in core Q# functionalities are using generic typing. Consider the function `Tail<'A> (array : 'A[])` from the `Microsoft.Quantum.Arrays` namespace, which returns the last element of an array. Because it is defined with a generic type parameter, the same function can be invoked against an array of any type. Listing 3.42 shows an example of this against an array of `Int` and `string`.

```
let labels = ["blue", "red", "green"];
let labelTail = Tail(labels); // "green"

let numbers = [1, 2, 3];
let numberTail = Tail(numbers); // 3
```
Listing 3.42 Invoking a generic function against two different types.

Type parameterization is allowed not only in the first level callable arguments, but also as part of a delegate being passed into the callable. Consider the built-in operation allowing to transform an array of one type to another, whose signature is shown in Listing 3.43.

```
operation ForEach<'T, 'U> (action : ('T => 'U),
    array : 'T[]) : 'U[]
```
Listing 3.43 A generic function with type parameterization inside the delegate parameter.

Not only does it contain two generic type parameters 'T and 'U, but also allows specifying a generic conversion operation 'T => 'U that will be applied on each of the array element.

3.3 Quantum Intermediate Representation

Q# programs do not compile to native quantum hardware instructions, but instead are intended to be compiled into *Quantum Intermediate Representation*. QIR is an LLVM based intermediate language which defines a hardware-agnostic set of rules for expressing the entire content of a quantum program. Different quantum hardware vendors can then choose to implement the entire QIR instruction set, or a subset of it, and translate this intermediate QIR program format into the hardware specific machine code. By relying on a single unified representation built on top of standard LLVM toolchain, it is much easier for various hardware platforms to execute quantum optimizers, transform the intermediate code and produce runnable machine code for specific execution targets.

Relying on QIR facilitates a smooth interweaving of classical and quantum computational logic in a single program, and allows the quantum hardware platforms to evolve easily, by progressively implementing larger sets of QIR instructions.

An example of a QIR representation of the simple Q# program from Listing 3.3 is shown in Listing 3.44. Despite looking a bit overwhelming, it is a fully self-contained description of the program. It contains definitions for all the types involved in the program's flow, the necessary constant declarations related to all hardcoded primitives and all functions corresponding to the capabilities of the program. All of those would have to be available on the given hardware platform for our program to be available with that quantum hardware.

```
%Range = type { i64, i64, i64 }
%String = type opaque

@PauliI = internal constant i2 0
@PauliX = internal constant i2 1
@PauliY = internal constant i2 -1
@PauliZ = internal constant i2 -2
@EmptyRange = internal constant %Range { i64 0, i64 1, i64 -1 }
@0 = internal constant [13 x i8] c"Hello world!\00"
@1 = internal constant [3 x i8] c"()\00"

define internal void @Basic__Main__body() {
entry:
  %0 = call %String* @__quantum__rt__string_create(i8*
      getelementptr inbounds ([13 x i8], [13 x i8]* @0, i32 0,
      i32 0))
  call void @__quantum__rt__message(%String* %0)
  call void @__quantum__rt__string_update_reference_count(%String
      * %0, i32 -1)
  ret void
}

declare %String* @__quantum__rt__string_create(i8*)

declare void @__quantum__rt__message(%String*)
```

```
declare void @__quantum__rt__string_update_reference_count(%
    String*, i32)

define void @Basic__Main__Interop() #0 {
entry:
  call void @Basic__Main__body()
  ret void
}

define void @Basic__Main() #1 {
entry:
  call void @Basic__Main__body()
  %0 = call %String* @__quantum__rt__string_create(i8*
      getelementptr inbounds ([3 x i8], [3 x i8]* @1, i32 0, i32
      0))
  call void @__quantum__rt__message(%String* %0)
  call void @__quantum__rt__string_update_reference_count(%String
      * %0, i32 -1)
  ret void
}

attributes #0 = { "InteropFriendly" }
attributes #1 = { "EntryPoint" }
```

Listing 3.44 A simple Q# program compiled into QIR.

Quantum Intermediate Representation itself is not coupled to Q#—in fact, quite the opposite, QIR is designed specifically in a way that any other programming language could support compiling its source code to a QIR-based format. QIR was initially developed by Microsoft as an integral feature of the Q# compiler and has since grown to become an independent project under the Linux Foundation, called the QIR Alliance,[4] formed at the end of 2021 [3]. At the time of writing the QIR Alliance steering members consist of representatives of Microsoft, Oak Ridge National Laboratory, Quantinuum, Quantum Circuits Inc., and Rigetti Computing.

3.4   Azure Quantum

The development of the QIR standard makes Q# programs ubiquitous, and potentially compatible with any quantum computing platform, operated by any hardware provider or cloud vendor. At the same time, Azure Quantum, being built by Microsoft, is the natural first choice cloud environment to deploy, execute and analyze quantum applications built with the QDK.

At the time of writing the Azure Quantum ecosystem offers public access to small integrated quantum computing devices from third-party vendors IonQ and Quantinuum, though this landscape is changing rapidly, with other partners such as Rigetti Quantum Cloud Services and Quantum Circuits, Inc. expected to launch

---

[4] https://qir-alliance.org.

their hardware on the platform soon. The tooling around Azure Quantum is under active development, and pricing models are still being defined [1]. Because of that general volatility around it, we specifically avoid discussing concrete integration or deployment into Azure Quantum on the pages of this study. Instead, the reader is recommended to follow official documentation for the latest developments in that space [7].

3.5 Quantum Simulators

3.5.1 Full State Simulator

The default simulator configured to run Q# programs when using the QDK is the QuantumSimulator, often referred to as *full state simulator*. It provides an accurate classical simulation of the quantum state of the program according to the mathematical formalism of quantum mechanics and the paradigms of quantum information theory.

It is idealized, and therefore it does not incorporate any randomized environment noise or fault conditions which would affect computations on physical quantum hardware. From that perspective, it always produces exact results or probability distributions as predicted by theoretical model.

Due to the exponential growth of the dimensions of Hilbert space with every additional qubit, the simulator is limited in the amount of qubits it can support—it can only handle 30 of them. Another obvious consequence of this is that the simulator is very resource intensive, and by default tries to use all available threads on the development machine. It also gets visibly slower with each extra qubit that is used in the program, especially once the qubit count reaches double digits.

A Q# program is executed from the command line on the full state simulator by performing the regular program execution using .NET SDK and by passing QuantumSimulator as the simulator parameter.

```
dotnet run --simulator QuantumSimulator
```
Listing 3.45 Invoking the Q# full state simulator from the command line.

In fact, because the full state simulator is the default one, omitting it completely when executing the program from the command line still makes the QDK fall back to it. The same behavior can be observed in other Q# environments, such as Jupyter Notebooks, where the IQ# magic command %simulate also ends up using the default full state simulator.

3.5.2 Toffoli Simulator

The Toffoli simulator is a unique QDK feature, allowing the full simulation a Q#
program, but only provided it is implemented using a small limited set of quantum
transformations—the **CNOT** and both single and multi-qubit controlled **X**.[5] Because
of such constraints, Toffoli simulator is able to simulate programs vastly exceeding
the total qubit capabilities of the full state simulator.

The simulator can be used from the command line using the Toffoli
Simulator name and in the Jupyter Notebooks environment it has a dedicated
command %toffoli.

```
dotnet run --simulator ToffoliSimulator
```
Listing 3.46 Invoking the Q# Toffoli simulator from the command line.

3.5.3 Resources Estimation

Another useful simulator shipping in the box with Quantum Development Kit is the
resources estimator. It is not a fully-fledged simulator in the strict meaning of the
word, because it does not fully simulate the Q# application. Instead, it analyzes the
program by statically detecting any qubit allocations and their releases, as well as
all quantum transformations enacted on those qubits—without actually attempting
to execute any of them. It then aggregates all this information into the summarized
quantum metric report, which is then returned as the output.

From that perspective, resources estimator can be used to analyze quantum programs far exceeding the simulation capabilities of the full state simulator, and, for the
time being, exceeding the execution capabilities of even the most advanced quantum
hardware.

In particular, the *width* and *depth* metrics of the program, reported by the estimator,
are useful information for the quantum programmers and researchers. Width indicates
how many qubits are needed for the program to execute, while depth provides an
estimated program execution time.

The estimator is available in the QDK under the ResourcesEstimator name. In
the Jupyter Notebooks, it can be used using the IQ# magic command %estimate.

```
dotnet run --simulator ResourcesEstimator
```
Listing 3.47 Invoking the Q# resources estimator from the command line.

The resources estimator is built on a lower level simulator component called
QCTraceSimulator. It is not accessible from the command line or Jupyter Notebook
environments, but can be used programmatically from C# or Python, and pointed at
a Q# program. It provides a superset of functionalities of the resources estimator and

---

[5] We will cover them in Chap. 4.

can produce more accurate information—at the expense of being more complicated to interact with.

References

1. Blog, Microsoft Q#. (2022). *Explore quantum hardware for free with azure quantum.* https://devblogs.microsoft.com/qsharp/explore-quantum-hardware-for-free-with-azure-quantum/. Accessed Feb 08 2022.
2. Filip, W. (2021). *Partial application in* Q#. https://qsharp.community/blog/partial-function-application/. Accessed Jan 03 2022.
3. Foundation, The Linux. (2021). *New quantum intermediate representation alliance serves as common interface for quantum computing development.* https://www.linuxfoundation.org/press-release/new-quantum-intermediate-representation-alliance-serves-as-common-interface-for-quantum-computing-development/. Accessed Dec 29 2021.
4. Kaiser, S. C., & Granade, C. (2021). *Learn quantum computing with python and Q#: A hands-on approach.* Manning.
5. Microsoft. (2021a). *LSP language specification.* https://microsoft.github.io/language-server-protocol/specifications/specification-current/. Accessed Jan 08 2022.
6. Microsoft. (2021b). *Q# language specification.* https://github.com/microsoft/qsharp-language/tree/main/Specifications/Language. Accessed Dec 30 2021.
7. Microsoft. (2022). *Azure quantum and Q# documentation.* https://docs.microsoft.com/en-us/azure/quantum/. Accessed Feb 07 2022.
8. Overflow, Stack. (2021). *2021 developer survey.* https://insights.stackoverflow.com/survey/2021. Accessed Jan 08 2022.

Part II

Chapter 4
Quantum Computing

In this chapter we are going to apply the learnings from Chap. 2 to the field of quantum computation. Most importantly, we shall cover the model of a qubit and its transformations, and draw parallels to the quantum mechanical theoretical concepts we covered earlier.

The reader will undoubtedly be pleasantly surprised to learn that our diligent study of two-level quantum systems was not in vain, because as it turns out, a qubit is nothing more than a generic name given to any such simplest quantum system. As such, we have actually been discussing qubits all along—without explicitly referring to them—and therefore we already covered all of the basic algebra needed to work with qubits in a quantum computer. Therefore, at least theoretically, any of the examples mentioned in Chap. 2, such as electron spin or photon polarization, could serve as an engine on top of which we could perform quantum computations. We will also explore the basic ways of constructing a quantum program using logical gates composed into quantum circuits.

Thanks to the time we spent discussing the QDK in Chap. 3, all of the content in this book will from now on be illustrated with tangible examples in Q#. By the end of the chapter we should be comfortable with the basic ideas behind building quantum programs, forming a solid foundation for the further chapters.

4.1 Qubits

In classical computing, the smallest indivisible unit of information is called a *bit*,[1] and can take two discrete values: 0 or 1. The entire information theory is built upon

---

[1] Some quantum computing literature refers to classical bits as *cbits*, to emphasize their classical nature. We shall not do that in this book though—we will assume that a term bit always refers to the classical concept.

© The Author(s), under exclusive license to Springer Nature Switzerland AG 2022
F. Wojcieszyn, *Introduction to Quantum Computing with Q# and QDK*, Quantum Science and Technology, https://doi.org/10.1007/978-3-030-99379-5_4

that, and our theoretical and practical models on how to transmit, process, store and encode information rely on the concept of bits.

On the other hand, in quantum information theory, the basic unit information is called a *qubit*, a *quantum* bit. Consequently, a classical computer performs its calculations by manipulating bits, while a quantum computer solves its tasks by operating on qubits.

A qubit, contrary to its less complicated classical cousin, could not only represent two binary states 0 or 1, but being a two-level quantum state, it can also be found in linear combination of them—forming a superposition of 0 and 1.[2] The fact that we can then perform computations over data in a superposition of those two states is one of the primary sources of potential advantages of quantum computing over the classical computation model. This is a phenomenon commonly referred to as *quantum parallelism*.

While we must be careful not to create an inaccurate mental picture, a simplified way of thinking about quantum parallelism is to consider that quantum computers can process evaluations of function $f(x)$ for many different values of x simultaneously [4]. As we will learn throughout this book, this effect is rather subtle, because as exciting and dramatic as quantum parallelism may sound to be, we are still limited by the measurement postulate of quantum mechanics. Therefore, we are only able to extract a single classical bit, 0 and 1, out of every qubit upon state measurement, so we should be careful when drawing any far reaching conclusions already.

While in the braket notation any labels assigned to the bras and kets are valid, it is a common convention in quantum information theory to use $|0\rangle$ and $|1\rangle$ as labels for the orthonormal basis corresponding to classical 0 and 1. This basis $\{|0\rangle, |1\rangle\}$ is referred to as the *computational basis* or the *standard basis*, and represents the reference in which all quantum algorithms are defined. The two linearly independent vectors making up the computational basis also happen to be the eigenvectors of Pauli's σ_z matrix.

$$|0\rangle = \begin{bmatrix} 1 \\ 0 \end{bmatrix} \qquad |1\rangle = \begin{bmatrix} 0 \\ 1 \end{bmatrix} \tag{4.1}$$

As a result, the computational basis is the same basis as we used in Chap. 2, where we denoted it as $\{|\uparrow\rangle, |\downarrow\rangle\}$ (see Eq. 2.10), hence it is also often called by the name we used already—the *Z-basis*.

The state of a single qubit, naturally in a two dimensional Hilbert space, would then be written in the standard basis as:

$$|\psi\rangle = a|0\rangle + b|1\rangle \tag{4.2}$$

---

[2] It is sometimes said that a qubit in a superposition "is both 0 and 1 at the same time". This is a rather simplistic and inaccurate description, but one that is commonly used in popular science articles, as it manages to convey the weirdness of conceptualizing quantum states. As we discussed earlier, it would be more appropriate to say that a qubit is neither 0 and 1, and only acquires a definite basis state upon measurement using a specific observable.

The second basis that is often used in quantum information theory is the basis composed out of the two eigenvectors of the Pauli's σ_x matrix—we already introduced it in Eq. 2.17 as the *X-basis*.[3] Its two orthogonal states are, by quantum computing convention, labeled $|+\rangle$ and $|-\rangle$.

$$|+\rangle = \frac{1}{\sqrt{2}}\begin{bmatrix}1\\1\end{bmatrix} = \frac{1}{\sqrt{2}}\Big(|0\rangle + |1\rangle\Big) \qquad |-\rangle = \frac{1}{\sqrt{2}}\begin{bmatrix}1\\-1\end{bmatrix} = \frac{1}{\sqrt{2}}\Big(|0\rangle - |1\rangle\Big) \quad (4.3)$$

In Q#, a use allocation block can be used to allocate a single qubit, several named qubits in a tuple or an array of qubits that can be referenced through an index. All three syntaxes are shown in Listing 4.1.

```
operation AllocateQubits() : Unit {
    use qubit = Qubit();
    use twoQubits = Qubit[2];
    use (qubit1, qubit2) = (Qubit(), Qubit());
    // all qubits can be interacted with
}
```
Listing 4.1 Different ways of allocating a qubit in Q#.

The scope for using the qubits in the Q# code is by default limited to the scope of the surrounding block, and in the case of Listing 4.1 that is the outer operation AllocateQubits (). Qubit allocations, however, can also be given their own explicit limited scope by creating a block allocation statement using curly braces.

```
operation AllocateQubits() : Unit {
    use qubit = Qubit() {
        // qubit scope is limited to this block
    }
    use twoQubits = Qubit[2] {
        // twoQubit array scope is limited to this block
    }
    use (qubit1, qubit2) = (Qubit(), Qubit()) {
        // qubit1 and qubit2 scope is limited to this block
    }
}
```
Listing 4.2 Allocating qubits in Q# with a dedicated scope block.

Our first interaction with the qubits in a more structured form will be building a small program that will allocate the qubits in their default state, measure them and keep track of the results. We will want to repeat the run for a larger number of iterations, because we are wary of the probabilistic nature of the measurements in quantum mechanics.

---

[3] An alternative name often used in literature is the *Hadamard basis*, named like this after a French mathematician, Jacques Hadamard

The core measurement functionalities in Q# can be found under the Microsoft.Quantum.Intrinsic namespace, in the form of M (qubit : Qubit) and Measure (bases : Pauli[], qubits : Qubit[]) operations. The former measures a single qubit in the default Z-basis—using the σ_z observable, while the latter is generalized operation which can measure any number of qubits in any of the three Pauli bases—X, Y or Z, corresponding to σ_x, σ_y and σ_z observables. Additional, more specialized, measurement operations are provided through the Microsoft.Quantum.Measurement namespace. The Pauli type makes it possible to use the measurement basis as a dynamic argument which can provide a lot of flexibility when authoring algorithms.

One thing worth noting at this point, is that contrary to some other quantum frameworks, Q# does not have any sampling and experiment repetition features. Therefore, in order to be able to measure the outcomes of our quantum operations in a larger sample size, the so-called "shots", we will need to perform the orchestration manually. On the other hand, because the language possesses all the usual constructs that are needed to facilitate such orchestration—arithmetic operations, loops or branching capabilities, it is not a huge obstacle to overcome. Just like it is the case for Pauli, Q# has keyword literals that are used when interpreting measurement results, corresponding to the eigenvalues of the selected observable. Since, as we know, those can be labeled anything, they are conveniently named Zero or One. They can be converted to integer or boolean representation if needed and can be used to handle measurement results in a strongly-typed manner.

A generic Q# helper operation that can be used for such sampling is shown below.

```
operation Sample(iterations : Int, basis : Pauli, op: (Pauli =>
    Result)) : Unit {
    mutable runningTotal = 0;
    for idx in 1..iterations {
        let result = op(basis);
        set runningTotal += result == One ? 1 | 0;
    }

    Message($"Measurement results:");
    Message($"|0>: {iterations - runningTotal}");
    Message($"|1>: {runningTotal}");
}
```

Listing 4.3 Helper Q# operation which we will use for testing purposes.

This orchestration operation allows the caller to pass in both the number of iterations and the intended measurement basis as arguments. In addition, the actual logical operation being tested is also passed in as a delegate. The orchestrator will invoke this delegate in a loop, capture its result and keep track of the total amount of 0 and 1 produced by it. This is done by comparing the measurement result, represented by the Result type, to the keyword literal One. If at the end of the entire routine runningTotal = 0, we would know that the measurements produced zeros only, if runningTotal == iterations we must have received only ones. Any total in between would indicate that there was some randomness in the measurement process.

Next, we need to define a simple Q# operation which is allocating a qubit in a default state and then measuring it in the relevant basis. This will be the operation that can be passed as delegate into the orchestration code.

```
operation MeasureDefaultQubit(basis : Pauli) : Result {
    use qubit = Qubit();
    let result = Measure([basis], [qubit]);
    return result;
}
```
Listing 4.4 Measuring a single qubit in Q#.

Such setup allows verifying two things—whether the newly allocated qubits really are guaranteed to have the state $|0\rangle$, and whether the measurements in the different Pauli bases really produce different results. As we recall from Chap. 2, superposition is basis dependent so a quantum state in an eigenstate with respect to one basis (e.g. Z-basis), will be in a superposition with respect to another basis (e.g. X-basis).

In order to find answers to these burning questions, we need to invoke the operation from 4.4 twice—once passing in PauliZ and then PauliX as the measurement basis.

```
@EntryPoint()
operation Main() : Unit {
    Message("Measuring default qubits in Z basis");
    Sample(4096, PauliZ, MeasureDefaultQubit);

    Message("Measuring default qubits in X basis");
    Sample(4096, PauliX, MeasureDefaultQubit);
}
```
Listing 4.5 Measuring qubits in different bases using the helper from Listing 4.3.

The produced output should resemble the following:

```
Measuring default qubits in Z basis
Measurement results:
|0>: 4096
|1>: 0

Measuring default qubits in X basis
Measurement results:
|0>: 2050
|1>: 2046
```
Listing 4.6 Expected output from the code from Listing 4.5.

When measuring in the Z-basis, we got 4096 zeros in 4096 iterations, which is quite encouraging. This is exactly what we expected, because Q# promises that newly initialized qubits have a $|0\rangle$ state, and the output of the program seems to agree. We also never applied any transformations to the qubit, so there is no reason to expect any deviations from $|0\rangle$.[4] The qubit started in an eigenstate of σ_z and since it never changed, the measurement result must be deterministic too.

On the other hand, when measuring in the X-basis, we experience maximum uncertainty. The distribution is not perfect, because the sample size is not very large,

---

[4] We are explicitly ignoring the topic of quantum hardware errors here.

but the conclusion that can be drawn of this is abundantly clear. When measuring in the X-basis, using σ_x observable, the measurement results are random—the state collapses to $|0\rangle$ or $|1\rangle$ with equal probability. As a result, the measurement produces each of 0 or 1 half of the time.

Q# imposes a requirement on the developers that at the end of each qubit scope— be it the surrounding implicit scope or the explicitly defined one—the qubit must either have been measured or manually reset to state $|0\rangle$.

In other words, while in its quantum memory management Q# does support automatic releasing of qubits at the end of an allocation scope, the released qubit must not be in a indeterminate state. In those situations manual reset is mandatory, and if the qubit was not measured or returned to state $|0\rangle$, a runtime (not compile time!) error would occur. On some quantum processors, a program state corruption may happen. Manual reset of dirty qubits can be done through the Reset (qubit : Qubit) operation for a single qubit, or through ResetAll (qubits : Qubit[]) for a larger number of qubits at once—in either case the qubits go back to $|0\rangle$ state. Listing 4.7 shows a correct reset procedure of qubits.

```
operation AllocateQubits() : Unit {
    use qubit = Qubit();
    // ... no measurement
    Reset(qubit);

    use twoQubits = Qubit[2];
    // ... no measurement
    ResetAll(twoQubits);

    use (qubit1, qubit2) = (Qubit(), Qubit());
    // ... no measurement
    ResetAll([qubit1, qubit2]);
}
```

Listing 4.7 Manually resetting qubits in Q#.

4.2   Quantum Circuits

In quantum computing, quantum state transformations are represented—similarly to the analogous concept from classical computer science—by computational gates called *quantum gates*. Such gates can then be composed into complex algorithmic sequences forming *quantum circuits*. Each gate is a unitary transformation that can be applied to the quantum state vector, and the size of the gate corresponds the amount of qubits it acts upon. Single qubit gates act on two-dimensional quantum state vectors and are thus described by 2×2 matrices. As a consequence of the tensor product representation of composite quantum states, two qubit gates act on four-dimensional quantum state vectors and are described by 4×4 matrices, three qubit gates act on 8 dimensional state vectors and are described by 8×8 matrices and so on. In general we can say that the gate matrix for n qubits has a $2^n \times 2^n$ matrix form.

Fig. 4.1 Sample quantum circuit

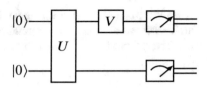

The primary role of circuits is to visualize the computation models as a sequence of qubit state transformations. The circuits provide a consistent graphical notation and a set of labeling conventions for expressing the flow of an algorithm, and they can accommodate operations on single qubits, as well as those transformations that are spanning multiple qubits.

The analogy between classical and quantum computing circuits works reasonably well, but a work of caution is in order. In classical computing, the gates typically correspond (albeit in a slightly idealized form) to physical electronic implementations. That is not necessarily the case in quantum computing. Depending on the implementation model of the physical quantum computer, a gate may really have a physical realization or they might be a purely logical concept, enacted on a stationary quantum object via external control device (e.g. a series of pulses in nuclear magnetic resonance or trapped ion quantum computer architectures) [5].

There are several useful tools that help authoring quantum circuits, two of the most popular being [5]Quirk, an online tool, and [6]quantikz, a LaTeX package.

A sample circuit is shown in Fig. 4.1 and describes a quantum algorithm that operates on two qubits, both of which are initialized to state $|0\rangle$. Each qubit in a quantum circuit is typically represented by an individual horizontal line, though as we will learn in later examples, it is also possible to group multiple qubits on a single line. The algorithm defines two quantum state transformations, first a **U** gate which spans both qubits, and then **V** gate applied only to the first qubit. Gates are normally visualized as squares or rectangles, with the appropriate letter corresponding to the transformation being executed. Finally, a measurement is performed on both of the qubits—which is indicated by the measurement symbols on both lines. Quantum circuits in general have no notion of measurements in bases different than computational basis—so the presence of the measurement symbol implies a σ_z observable. The circuit then changes both qubit lines into double lines—this indicates that at that point the qubit has been measured and carries a classical bit—0 or 1.

One thing that may be confusing is that quantum circuits are written and left to write, and therefore logically interpreted this way too. On the other hand, the mathematical description of matrix operations representing the same quantum algorithm is written right to left. The matrix equivalent of Fig. 4.1 is shown in Eq. 4.4 below.

$$(\mathbf{V} \otimes \mathbf{I})\mathbf{U} |00\rangle \tag{4.4}$$

---

[5] https://algassert.com/quirk.

[6] https://ctan.org/pkg/quantikz.

It should be noted that the usage of $\mathbf{I}$ matrix implies a no-op transformation on the qubit—something that on the circuit can also be expressed by a simple whitespace on a given qubit line.

4.3   H Gate and the Superposition

We already referred to *superposition* as being one of the primary sources of potential computational advantages for quantum computing, but so far we only interacted with qubits in their default state. Computing in a superposition state is utilized in just about any quantum algorithm, so being able to create a superposition state is one of the fundamental preparation steps in quantum computing.

In Chap. 2, in Eq. 2.43, we defined a simple matrix representation for a beam splitter that would create a uniform superposition of both of the possible photon states. As it turns out, the exact same transformation has profound importance in quantum information theory, and it is known as $\mathbf{H}$, the *Hadamard gate*.

$$\mathbf{H} = \frac{1}{\sqrt{2}} \begin{bmatrix} 1 & 1 \\ 1 & -1 \end{bmatrix} \tag{4.5}$$

The Hadamard transformation is extremely useful, because of its ability to create a uniform superposition out of any of the two computational basis states. Suppose we start with a qubit prepared to be $|0\rangle$. The $\mathbf{H}$ transformation then produces the following:

$$\mathbf{H} |0\rangle = \frac{1}{\sqrt{2}} |0\rangle + \frac{1}{\sqrt{2}} |1\rangle = \frac{1}{\sqrt{2}} \left(|0\rangle + |1\rangle \right) \tag{4.6}$$

Similarly, when starting in the $|1\rangle$ state, the effect of applying $\mathbf{H}$ is:

$$\mathbf{H} |1\rangle = \frac{1}{\sqrt{2}} |0\rangle - \frac{1}{\sqrt{2}} |1\rangle = \frac{1}{\sqrt{2}} \left(|0\rangle - |1\rangle \right) \tag{4.7}$$

As we know from the beam splitter discussion, the difference between the two situations is the amplitude sign ($(+)$ vs $(-)$), which indicates the *phase* difference between the states. At the same time, according to the Born rule, in order to obtain classic probabilities we need to square the amplitudes, we are equally likely to have the state collapse to $|0\rangle$ and to $|1\rangle$ upon measurement. In other words, the Hadamard transformation always creates a uniformly distributed superposition, irrespective of the initial state of the qubit, as long as that state was one of the computational basis states $\{|0\rangle, |1\rangle\}$.

The Hadamard matrix is not just unitary, it is also self adjoint (Hermitian), as it is equal to its own conjugate transpose. In other words:

$$\mathbf{H} = \mathbf{H}^{\dagger} \tag{4.8}$$

This means that the following holds:

$$\mathbf{HH^\dagger} = \mathbf{HH} = \mathbf{I} \tag{4.9}$$

The obvious consequence of this is that when applying the Hadamard transformation to a qubit twice in a row, the second transformation would undo the effects of the first one and restore the original quantum state. For example:

$$\mathbf{H}\frac{1}{\sqrt{2}}\left(|0\rangle + |1\rangle\right) = |0\rangle \tag{4.10}$$

The effect that manifests itself here is the interference again, the same as we saw in Sect. 2.4. The Hadamard gate causes constructive interference of the probability amplitudes related to $|0\rangle$, which becomes amplified in a way that we are now certain to measure $|0\rangle$. Conversely, destructive interference of the amplitudes on state $|1\rangle$ results in the probability of measuring it dropping to 0.

An elegant way of expressing the Hadamard gate acting on a single qubit in a computational basis state is:

$$\mathbf{H}|x\rangle = \frac{1}{\sqrt{2}}\left(|0\rangle + (-1)^x |1\rangle\right) \tag{4.11}$$

where $x \in 0, 1$. This captures the phase change that happens depending on whether we apply $\mathbf{H}$ to $|0\rangle$ or $|1\rangle$. We can further rewrite this into a format relevant for single qubit input $|x\rangle$ in any superposition state:

$$\mathbf{H}|x\rangle = \frac{1}{\sqrt{2}} \sum_{z \in \{0,1\}} (-1)^{xz} |z\rangle \tag{4.12}$$

Finally, there is one other alternative way of expressing the effect of the $\mathbf{H}$ gate on a single computational basis state qubit, using Euler's number e:

$$\mathbf{H}|x\rangle = \frac{1}{\sqrt{2}}\left(|0\rangle + e^{\pi i x} |1\rangle\right) \tag{4.13}$$

where $x \in 0, 1$. This works because the exponent can evaluate to 1 or -1 only, depending on the value of x.

Another interesting point of view to consider when dealing with the Hadamard gate is that we can view the Hadamard transformation as a mean to change the basis—from Z-basis to X-basis, and vice versa. The reason for that is that the states from Eqs. 4.6 and 4.7 actually corresponds to the basis states of the X-basis $\{|+\rangle, |-\rangle\}$.

$$\mathbf{H}|0\rangle = |+\rangle \qquad \mathbf{H}|1\rangle = |-\rangle \tag{4.14}$$

And of course, since **H** is Hermitian and thus its own inverse, the following holds as well:

$$\mathbf{H}\,|+\rangle = |0\rangle \qquad \mathbf{H}\,|-\rangle = |1\rangle \tag{4.15}$$

While the Hadamard gate is a single qubit transformation, it is naturally possible to use it in multi-qubit systems as well—in those cases it needs to be applied to the individual qubits separately. Let us now consider a two qubit system and see what happens when apply the **H** transformation to each of the individual qubits that make it up. Since we already know that a multi-qubit system is described fully by a tensor product of the individual component states, we can write this as follows, assuming we start with two qubits in state $|0\rangle$ each.

$$\mathbf{H}\,|0\rangle \otimes \mathbf{H}\,|0\rangle = \left(\frac{1}{\sqrt{2}}\,|0\rangle + \frac{1}{\sqrt{2}}\,|1\rangle \right) \otimes \left(\frac{1}{\sqrt{2}}\,|0\rangle + \frac{1}{\sqrt{2}}\,|1\rangle \right) =$$
$$\frac{1}{2}\left(|00\rangle + |01\rangle + |10\rangle + |11\rangle \right) \tag{4.16}$$

It is possible to perform a similar calculation for the other three possible combinations of the computational basis states. The difference between the results will be, just like in the case of a single qubit **H**, the phase.

$$\mathbf{H}\,|0\rangle \otimes \mathbf{H}\,|1\rangle = \left(\frac{1}{\sqrt{2}}\,|0\rangle + \frac{1}{\sqrt{2}}\,|1\rangle \right) \otimes \left(\frac{1}{\sqrt{2}}\,|0\rangle - \frac{1}{\sqrt{2}}\,|1\rangle \right) =$$
$$\frac{1}{2}\left(|00\rangle - |01\rangle + |10\rangle - |11\rangle \right) \tag{4.17}$$

$$\mathbf{H}\,|1\rangle \otimes \mathbf{H}\,|0\rangle = \left(\frac{1}{\sqrt{2}}\,|0\rangle - \frac{1}{\sqrt{2}}\,|1\rangle \right) \otimes \left(\frac{1}{\sqrt{2}}\,|0\rangle + \frac{1}{\sqrt{2}}\,|1\rangle \right) =$$
$$\frac{1}{2}\left(|00\rangle + |01\rangle - |10\rangle - |11\rangle \right) \tag{4.18}$$

$$\mathbf{H}\,|1\rangle \otimes \mathbf{H}\,|1\rangle = \left(\frac{1}{\sqrt{2}}\,|0\rangle - \frac{1}{\sqrt{2}}\,|1\rangle \right) \otimes \left(\frac{1}{\sqrt{2}}\,|0\rangle - \frac{1}{\sqrt{2}}\,|1\rangle \right) =$$
$$\frac{1}{2}\left(|00\rangle - |01\rangle - |10\rangle + |11\rangle \right) \tag{4.19}$$

In general, any multi-qubit transformation like this can be written in the tensor product form

$$(\mathbf{H} \otimes \mathbf{H} \otimes \ldots \otimes \mathbf{H})\,|00\ldots0\rangle \tag{4.20}$$

By utilizing the tensor product, we can describe the two-qubit Hadamard transformation in a form of matrix of matrices, namely a new $\mathbf{H}^{\otimes 2}$ matrix. That matrix itself is made up of **H** matrices.

$$\mathbf{H}^{\otimes 2} = \frac{1}{\sqrt{2}} \begin{bmatrix} \mathbf{H} & \mathbf{H} \\ \mathbf{H} & -\mathbf{H} \end{bmatrix} \tag{4.21}$$

This reasoning can be generalized even further, into a $\mathbf{H}^{\otimes n}$ transformation that recursively acts on an arbitrary number of n qubits.

$$\mathbf{H}^{\otimes n} = \frac{1}{\sqrt{2}} \begin{bmatrix} \mathbf{H}^{\otimes(n-1)} & \mathbf{H}^{\otimes(n-1)} \\ \mathbf{H}^{\otimes(n-1)} & -\mathbf{H}^{\otimes(n-1)} \end{bmatrix} \tag{4.22}$$

The transformation $\mathbf{H}^{\otimes n}$ is sometimes referred to as the so-called Walsh-Hadamard (or sometimes just Walsh) transformation $\mathbf{W}$, though in this account we shall continue using $\mathbf{H}^{\otimes n}$ as it is more verbose. We can summarize the effects of a $\mathbf{H}^{\otimes n}$ multi-qubit transformation on an input state $|00..0\rangle$ (n qubits in $|0\rangle$ state each) as:

$$\mathbf{H}^{\otimes n} |00..0\rangle = \frac{1}{\sqrt{2^n}} \sum_{x=0}^{2^n-1} |x\rangle \tag{4.23}$$

While this provides a nice generalization for the case of using n inputs, they are all still restricted to being in $|0\rangle$ initial state only. We can now combine Eqs. 4.12 and 4.23. We can assume that $|x\rangle$ in Eq. 4.12 has a length of n, instead of being a single qubit, it leads us to the final generalization of the $\mathbf{H}^{\otimes n}$ transformation. The formula is as follows

$$\mathbf{H}^{\otimes n} |x\rangle = \frac{1}{\sqrt{2^n}} \sum_{z=0}^{2^n-1} (-1)^{x \cdot z} |z\rangle \tag{4.24}$$

where $x \cdot z$ is the bitwise inner product of x and z, modulo 2 (binary scalar product)

$$x \cdot z = x_0 z_0 \oplus x_1 z_1 \oplus \ldots \oplus x_{n-1} z_{n-1} \tag{4.25}$$

With all this theory under our belt, we are now well positioned to take the simple default qubit measurement examples from the previous section and extend them to involve the usage of superposition created with the Hadamard transformation.

The first basic program will implement the circuit defined below. The newly allocated qubit has the definite state $|0\rangle$, the **H** transformation is applied and the measurement follows.

In Q# the **H** transformation is exposed as H (qubit : Qubit) operation in the Microsoft.Quantum.Intrinsic namespace. As we did in Listing 4.4, it would

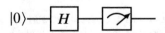

Fig. 4.2 Creating a uniform superposition with the **H** transformation

make sense to have the possibility for measuring in different bases, in order to be able to verify the impact of different bases on this uniform superposition (at this point it should be clear that superposition in one basis is not necessarily a superposition in another).

```
operation MeasureSuperpositionQubit(basis : Pauli) : Result {
    use qubit = Qubit();
    H(qubit);
    let result = Measure([basis], [qubit]);
    return result;
}
```

Listing 4.8   Q# code applying the **H** gate, and supporting measurements in an arbitrary basis.

Just as before, we should run this circuit a larger number of times to gather certain representative set of results. The code will therefore reuse the sampling orchestrator from Listing 4.3—the operation can be invoked for both Z and X-bases using the following code.

```
@EntryPoint()
operation Main() : Unit {
    Message("Measuring Z-basis superposition in the Z basis");
    Sample(4096, PauliZ, MeasureSuperpositionQubit);

    Message("Measuring Z-basis superposition in the X basis");
    Sample(4096, PauliX, MeasureSuperpositionQubit);
}
```

Listing 4.9   Orchestration of the test code from Listing 4.8.

Of course based on the fact that the **H** transformation is effectively a basis switch between Z- and X-bases, which is best shown in Eqs. 4.14 and 4.15, the expected result is that measurements in the Z-basis produce evenly distributed 0 and 1 results, while the X-basis measurements produce only 0.

```
Measuring Z-basis superposition in the Z basis
Measurement results:
|0>: 2041
|1>: 2055

Measuring Z-basis superposition in the X basis
Measurement results:
|0>: 4096
|1>: 0
```

Listing 4.10   Sample output from running the code from Listing 4.9.

This is obviously a rather modest example of a quantum program, but the consequences are quite astonishing. Thanks to the Hadamard transformation, we managed to generate a truly nature-guaranteed random bit.[7] The qubit in the program had 50%

---

[7] The reader may point out that we already achieved that when measuring a default in the X-basis in the previous section. This is a correct observation—however it is the computational basis that is really the standard way of expressing quantum algorithms, and indeed in this particular example a measurement is a standard Z-basis measurement.

chance of collapsing to 0 or 1 each, and those probabilities are guaranteed by the laws of quantum mechanics and its underlying mathematics. This is something that one cannot achieve in classical computing—where a truly random bit generation is impossible.

Of course for most problems, a meager single bit is not a particularly useful computation result. What we can do, however, is extend the random bit generation onto a set of bits, which will then allow us to produce a random bit array, which can be used to express more sophisticated information, such as for example a random integer.

Let us assume we would want to generate $n = 8$ random bits, to produce a random unsigned 8-bit integer. There are three ways that we will approach this. First, a most naive example will see us simply repeating the procedure of measuring a single qubit n times, until we have all the necessary bits. With such approach, we still operate on one individual qubit only and the circuit design from Fig. 4.2 still applies—it will however get repeated n amount of times.

```
operation RandomNumber1Qubit(bitCount : Int) : Int {
    mutable randomBits = [false, size = bitCount]; // 000...0

    for idx in 0..bitCount-1 {
        use qubit = Qubit();
        H(qubit);
        let result = M(qubit);
        set randomBits w/= idx <- result == One;
    }

    return BoolArrayAsInt(randomBits);
}
```

Listing 4.11 A sample Q# random number generator using a single qubit.

The solution is shown in Listing 4.11 and initially allocates a mutable array of bits, by default set to all zeroes. Then, as the code performs measurements of the individual qubit results in a loop, if the measurement produces a 1, an update, through Q#'s copy-and-update expression, is performed on the array index corresponding to the qubit being iterated over.

A more elaborate solution to the same problem, is to use all the required qubits simultaneously. This of course might be problematic for a really large number of bits that are required, given the quantum hardware constraints today, but for the sake of an example let us see how such a solution would look like.

In order to create a superposition over eight qubits, we will rely on the $\mathbf{H}^{\otimes 8}$ transformation, which can be represented on a quantum circuit in a compact format using the annotations shown in Fig. 4.3.

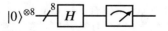

Fig. 4.3 Creating a uniform superposition with the **H** transformation over 8 qubits

Q# provides two additional framework features that come in particularly useful when implementing this particular quantum circuit. First of all, the ApplyToEach<'T> (singleElementOperation : ('T => Unit), register : 'T[]) operation, from the Microsoft.Quantum.Canon namespace, allows invoking a specific operation delegate against an array of qubits. This provides a rather concise way of preparing or transforming multi-qubit states. Secondly, a multi-qubit measurement helper MultiM (targets : Qubit[]), from the Microsoft.Quantum.Measurement namespace, allows a convenient measurement of an array of qubits in the computational basis, and returns an array of Result objects back, with the index of the Result in the output array corresponding to the index the qubit occupied in the measurement input array. Both of these are convenience features that help keep the amount of the involved Q# code to the bare minimum—something that is rather welcome by software engineers. The entire operation is shown in Listing 4.12.

```
operation RandomNumberNQubits(bitCount : Int) : Int {
    use qubits = Qubit[bitCount];
    ApplyToEach(H, qubits);

    let result = MultiM(qubits);
    return BoolArrayAsInt(ResultArrayAsBoolArray(result));
}
```

Listing 4.12 A sample Q# random number generator using n qubits and manual bit sampling.

The third approach to this problem, takes an even bigger dependency on the built-in Q# functionalities. The particular feature that we take advantage of here, is the ability to create a so-called QPU register, a LittleEndian register, which is a built-in Q# user defined type under the Microsoft.Quantum.Arithmetic namespace. The endianness is the classical computing concept indicating the sequencing of bits in a byte—a little-endian approach stores the least-significant bit at the smallest address. Q# supports both big endian and little endian registers, and can create them from an array of qubits in order to simplify arithmetic operations.

We can then measure the register directly into an integer using yet another measurement helper operation—MeasureInteger (target : LittleEndian) operation from the same Microsoft.Quantum.Arithmetic namespace. This measurement extension simplifies the process of performing arithmetic calculations on the QPU, as one no longer has to manually handle the raw result bits. The entire code of this third version of the quantum random number generator in Q# is shown in Listing 4.12.

```
operation RandomNumberFramework(bitCount : Int) : Int {
    use qubits = Qubit[bitCount];
    ApplyToEach(H, qubits);

    let register = LittleEndian(qubits);
    let randomNumber = MeasureInteger(register);
    return randomNumber;
}
```

Listing 4.13 A sample Q# random number generator using n qubits and the built-in standard readout of integers.

Irrespective of the approach taken, each of the three random number generation operations from examples Listing 4.11, 4.12 and 4.13, can be invoked in a similar fashion

```
@EntryPoint()
operation Main() : Unit {
    let rnd1 = RandomNumber1Qubit(8);
    Message($"Random uint8 using single qubit: {rnd1}");

    let rnd2 = RandomNumberNQubits(8);
    Message($"Random uint8 using multiple qubits: {rnd2}");

    let rnd3 = RandomNumberFramework(8);
    Message($"Random uint8 using framework features: {rnd3}");
}
```
Listing 4.14 Orchestration code to test the random number generators.

And the expected output is going to be three truly random[8] 8-bit unsigned integers. For example:

```
Random uint8 using single qubit: 32
Random uint8 using multiple qubits: 176
Random uint8 using framework features: 179
```
Listing 4.15 Sample output from running the code from Listing 4.14.

4.4 Pauli Gates

Four of the most basic single qubit gates are represented by the Pauli matrices σ_i, σ_x, σ_y and σ_z. Despite the fact that in the mathematical formalism of quantum mechanics Pauli matrices are predominantly utilized as observables in the measurement process, in quantum computing, due to their interesting algebraic properties, they are also useful as regular unitary state transformations. Together they form a set of some of the most basic qubit operations that one can execute in quantum programs. These gates are customarily referred to as the **I**, **X**, **Y** and **Z** gates.

In this section, we will continue using the sampling code from Listing 4.3, however at this point we can simplify it slightly and get rid of the parameterized basis—from now on we are only going to be interested in computational basis measurements. The updated orchestrator code is therefore:

---

[8] We are obviously taking a simplified approach here. Quantum random number generation is a major pillar of cryptography and a deeply complex field. Readers willing to dive deeper into the related challenges and statistical consequences are referred to specialized studies in the topic such as [3].

```
operation Sample(iterations : Int, op: (Unit => Result)) :Unit {
    mutable runningTotal = 0;
    for idx in 1..iterations {
        let result = op();
        set runningTotal += result == One ? 1 | 0;
    }

    Message($"Measurement results:");
    Message($"|0>: {iterations - runningTotal}");
    Message($"|1>: {runningTotal}");
}
```

Listing 4.16 Adjusted orchestration code from Listing 4.3.

4.4.1 I Gate

Identity gate **I**, sometimes denoted as **1** as well, is the simplest possible quantum gate. When applied to any quantum state vector, **I** leaves the state intact. Its matrix representation is the same as σ_i:

$$\mathbf{I} = \begin{bmatrix} 1 & 0 \\ 0 & 1 \end{bmatrix} \tag{4.26}$$

The identity gate can be interpreted similarly to identity gate in classical computing, as an identity function, namely a function that takes in an argument and returns that argument intact (Fig. 4.4).

$$\mathbf{I}\,|\psi\rangle = |\psi\rangle \tag{4.27}$$

Of course the effect of the identity gate on an arbitrary qubit in a computational basis superposition is equally unremarkable

$$\mathbf{I}(a\,|0\rangle + b\,|1\rangle) = a\,|0\rangle + b\,|1\rangle \tag{4.28}$$

The main use case of identity is to indicate that no state transformation is performed on a given qubit in a multi-qubit composite system. This is something we mentioned already in Sect. 4.3. Another useful aspect of **I** is that it can often be used to express various properties of other gates. For example, if the following condition holds for an unknown gate **U**:

$$\mathbf{I} = \mathbf{U}\mathbf{U} \tag{4.29}$$

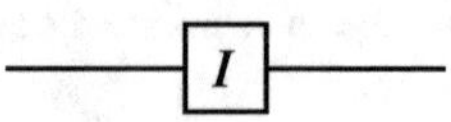

Fig. 4.4 Identity gate

we know that the gate **U** is self-adjoint—Hermitian.

Although in itself it is not particularly useful, for the sake of completeness we shall mention here that Q# also ships with an implementation of the **I** gate. It is part of the Microsoft.Quantum.Intrinsic namespace and exposed as a I (target : Qubit) no-op operation. Given that Q# is so easy to write and test, we can verify that by running a larger sample set of an operation applying **I** to the default |0⟩ and measuring the qubit afterwards. This is very similar to the sample from Listing 4.8.

```
operation MeasureI() : Result {
    use qubit = Qubit();
    I(qubit);
    let result = M(qubit);
    return result;
}
```

Listing 4.17 Q# code testing the behvior of the **I** gate.

Naturally, such operation produces all zeroes—this becomes evident once we invoke it a larger number of times using the orchestration code from Listing 4.16. In the particular example below, it was done 4096 times.

```
Measuring I
Measurement results:
Result 0: 4096
Result 1: 0
```

Listing 4.18 Output from the repeated orchstration of code from Listing 4.17.

4.4.2 X Gate

The Pauli **X** gate is often (Fig. 4.5) referred to as the *bit-flip gate*, because it swaps the probability amplitudes a and b with each other. Geometrically, the gate performs a rotation by π about the x-axis of the Bloch sphere. Its matrix representation is the same as σ_x

$$\mathbf{X} = \begin{bmatrix} 0 & 1 \\ 1 & 0 \end{bmatrix} \tag{4.30}$$

We can express the linear transformation it enacts on a qubit state as

$$\mathbf{X}(a\,|0\rangle + b\,|1\rangle) = b\,|0\rangle + a\,|1\rangle \tag{4.31}$$

If the qubit is already in a Z-basis eigenstate |0⟩ or |1⟩, in other words, we know that the probability of one of the two basis states is zero, then the gate acts like the NOT gate in classical computing—flipping the state vector to its orthogonal counterpart.

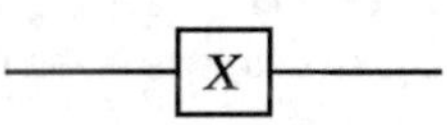

Fig. 4.5 Bit-flip gate

Fig. 4.6 Phase-flip gate

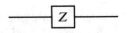

$$\mathbf{X}\,|0\rangle = |1\rangle \qquad \mathbf{X}\,|1\rangle = |0\rangle \tag{4.32}$$

This conceptual symmetry to the classical NOT gate makes the Pauli **X** gate one of the quantum gates that is easiest to understand. We can test the effects of the **X** gate by updating the code from Listing 4.17 to make use of the built-in Q# X (target : Qubit) operation.

```
operation MeasureX() : Result {
    use qubit = Qubit();
    X(qubit);
    let result = M(qubit);
    return result;
}
```

Listing 4.19 Q# code testing the behvior of the **X** gate.

Since the newly allocated qubits are guaranteed to be in state $|0\rangle$, the code from Listing 4.19 would produce only states equal to $|1\rangle$. We can verify that by running the repeated measurements using our orchestration code—again 4096 times:

```
Measuring X
Measurement results:
|0>: 0
|1>: 4096
```

Listing 4.20 Output from the repeated orchstration of code from Listing 4.19.

4.4.3   Z Gate

Contrary to the **X** gate, which can be relatively easily conceptually mapped to classical computation theory, the **Z** gate performs a transformation whose effects are uniquely quantum, making it a bit more difficult to apprehend. Mathematically, the gate is very simple—it is represented by the same matrix as σ_z and (Fig. 4.6) performs a rotation by π about the z-axis of the Bloch sphere.

$$\mathbf{Z} = \begin{bmatrix} 1 & 0 \\ 0 & -1 \end{bmatrix} \tag{4.33}$$

The effects on an arbitrary quantum state are thus

$$\mathbf{Z}(a\,|0\rangle + b\,|1\rangle) = a\,|0\rangle - b\,|1\rangle \tag{4.34}$$

This is of course the exact same thing that we saw when modelling the phase shifter when discussing the Mach-Zehnder interferometer in Sect. 2.4. As a consequence, the effects the **Z** gate has on qubits will be the same—namely, the only thing that the **Z** gate changes is the sign of the second amplitude, which flips from $(+)$ to $(-)$.

Because of that, the **Z** gate is commonly referred to as the *phase flip gate*, as it changes the *phase* of the qubit, leaving the actual classical probabilities intact. A simple way of remembering the effect of the **Z** gate on the Z-basis eigenstate is that it leaves $|0\rangle$ intact and multiplies $|1\rangle$ by -1.

$$\mathbf{Z}|0\rangle = |0\rangle \qquad \mathbf{Z}|1\rangle = -|1\rangle \tag{4.35}$$

A more concise way of expressing that notion would be

$$\mathbf{Z}|x\rangle = (-1)^x |x\rangle \tag{4.36}$$

where $x \in 0, 1$.

Despite having no impact on classical probabilities, the **Z** gate has fundamental importance and a wide array of useful application scenarios in quantum computing, many of which we will cover in this book. In particular, as an integral attribute of the wave-particle nature of quantum mechanism, phase differences can amplify each other, or cancel each other out via the interference effect—something that can be taken advantage of in quantum algorithms. Additionally, phase difference allows for flagging a specific quantum state with the relevant sign allowing us to differentiate between states that would otherwise be indistinguishable.

One other thing worth noting, is that while the probability amplitudes in the Z-basis do not get affected by the **Z** transformation, this is not the case in other bases. For example in the X-basis, the **Z** gate would flip $|+\rangle$ into $|-\rangle$ state.

Q#, when running on a simulator allows close inspection of the current quantum state of the program. During development, this is something that is very helpful in ensuring that the quantum state vector is really corresponding to the algebraic expectations that ones has. This functionality is provided by a function belonging to the very helpful `Microsoft.Quantum.Diagnostics` namespace, called `DumpMachine<'T>` (`location : 'T`), and can be used to output a simple visualization of the wave function describing the current system state into a desired location, such as an external file. It can also be invoked without any parameters in order to write out that state to console (standard I/O).

Because we cannot directly detect the effects of the **Z** gate using measurements, we can use this particular technique to inspect its impact instead. Listing 4.21 shows two single qubit operations that make use of the machine dump. the first one would show the effects of the **Z** gate on the default $|0\rangle$, while the second on the $|1\rangle$ basis state. At this point we already know that $|1\rangle$ can be obtained in code by first invoking the **X** transformation.

```
operation TestZAgainst0() : Unit {
    use qubit = Qubit();
    Z(qubit);
    DumpMachine();
    Reset(qubit);
}

operation TestZAgainst1() : Unit {
```

```
    use qubit = Qubit();
    X(qubit);
    Z(qubit);
    DumpMachine();
    Reset(qubit);
}
```
Listing 4.21 Q# code testing the behvior of the **Z** gate on a quantum simulator.

The operations can be invoked using the following code:

```
@EntryPoint()
operation Main() : Unit {
    Message("Testing Z against |0>");
    TestZAgainst0();

    Message("Testing Z against |1>");
    TestZAgainst1();
}
```
Listing 4.22 Test code orchestrating operations from Listing 4.21.

The expected output is then as follows:

```
Testing Z against |0>
# wave function for qubits with ids: 0
|0>:1.0000 + 0.0000i ==  **********  [1.0000]   --- [0.00000 rad]
|1>:0.0000 + 0.0000i ==              [0.0000]

Testing Z against |1>
# wave function for qubits with ids: 0
|0>: 0.0000 + 0.0000i ==             [0.0000]
|1>:-1.0000 + 0.0000i ==  **********  [1.0000] ---    [3.14159 rad]
```
Listing 4.23 Expected output from code from Listing 4.22.

A few words are required on how to read the machine dump. First of all, a qubit label using the little endian notation is shown in the left most column, corresponding to the possible states—it would be a range from $|0\rangle$ to $|2^n\rangle$ for n-qubit state machine. Then the probability amplitude is shown in a Cartesian format, with explicit real and imaginary components. In the particular example in Listing 4.23 the imaginary numbers are not needed to describe the quantum state so they are shown as 0. The asterisks visualize the approximated probability amplitude distributions between the possible measurement outcomes. Next, an accurate value of the probability amplitude as a decimal number is displayed, followed by a numeric value of the phase shown in radians. In the sample snippet above the phase equals to π because $e^{i\pi} = -1$, confirming our expectations for the transformation $Z|1\rangle$.

It must be emphasized again that such state debugging technique only works on simulators—when executing against real quantum hardware, all the quantum mechanical rules apply and thus a state can only be known with certainty after a measurement using a relevant observable. No intermediary state peeking is possible.

Fig. 4.7 Pauli **Y** gate

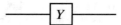

4.4.4 Y Gate

Pauli **Y** gate is *almost* a combination of both **X** and **Z** gates, as it effectively performs both a bit-flip and a phase-flip at the same time. Geometrically, the gate performs a rotation by π about the y-axis of the Bloch sphere (Fig. 4.7).

The matrix of **Y** form is the same as σ_y

$$\mathbf{Y} = \begin{bmatrix} 0 & -i \\ i & 0 \end{bmatrix} \tag{4.37}$$

When applied to the basis states $|0\rangle$ and $|1\rangle$, Y produces the following states

$$\mathbf{Y}|0\rangle = i|1\rangle \qquad \mathbf{Y}|1\rangle = -i|0\rangle \tag{4.38}$$

And when applied to a general superposition state

$$\mathbf{Y}(a|0\rangle + b|1\rangle) = -ib|0\rangle + ia|1\rangle = -i(b|0\rangle - a|1\rangle) \tag{4.39}$$

The actual relationship between the bit-flip of **X** and phase-flip of **Z** is therefore

$$\mathbf{XZ} = -i\mathbf{Y} \tag{4.40}$$

We can verify the detailed effects using the Q# diagnostic tools again. The code below performs quantum state dumps against two operations that invoke the **Y** transformation on input states of $|0\rangle$ and $|1\rangle$.

```
operation TestYAgainst0() : Unit {
    use qubit = Qubit();
    Y(qubit);
    DumpMachine();
    Reset(qubit);
}

operation TestYAgainst1() : Unit {
    use qubit = Qubit();
    X(qubit);
    Y(qubit);
    DumpMachine();
    Reset(qubit);
}
```

Listing 4.24 Q# code testing the behvior of the **Y** gate on a quantum simulator.

The output will show that indeed the state $|0\rangle$ was transformed to $i|1\rangle$ while $|1\rangle$ to $-i|0\rangle$.

```
Testing Y against |0>
# wave function for qubits with ids: 0
|0>:0.0000 + 0.0000i ==             [0.0000]
|1>:0.0000 + 1.0000i == ********** [1.0000]      ^      [1.57080 rad]

Testing Y against |1>
# wave function for qubits with ids: 0
|0>:0.0000 +-1.0000i == ********** [1.0000]      v      [-1.57080 rad]
|1>:0.0000 + 0.0000i ==             [0.0000]
```

Listing 4.25 Expected output of testing the **Y** gate on a quantum simulator.

As expected, in both cases the imaginary component of the probability amplitude was acquired—once as i and once as $-i$.

4.4.5 Further Considerations

Just as we mentioned in Sect. 2.5 when discussing observables, each of the three Pauli gates can also be written using Dirac notation, as the outer product of the Z-basis vectors with their complex conjugates.

$$\mathbf{I} = |0\rangle\langle0| + |1\rangle\langle1| \qquad \mathbf{X} = |1\rangle\langle0| + |0\rangle\langle1| \qquad (4.41)$$
$$\mathbf{Y} = -|1\rangle\langle0| + |0\rangle\langle1| \qquad \mathbf{Z} = |0\rangle\langle0| - |1\rangle\langle1|$$

The Pauli gates are all Hermitian and each one is its own inverse, meaning the following holds

$$\mathbf{X}^2 = \mathbf{Y}^2 = \mathbf{Z}^2 = \mathbf{I}^2 = -i\mathbf{XYZ} = \mathbf{I} \qquad (4.42)$$

In practice, it means that applying the gate the second time in a row would undo the effects of the first transformation. Additionally, Pauli gates are closely related to the Hadamard gate, as we can treat the **H** gate as "basis change"-gate. Therefore, all of the following relations are true

$$\mathbf{X} = \mathbf{HZH} \qquad (4.43)$$
$$\mathbf{Y} = -\mathbf{HYH}$$
$$\mathbf{Z} = \mathbf{HXH}$$

These relationships are easy enough to verify using linear algebra, but it is naturally very tempting to just experiment with such configurations in Q# code. One such example is shown in Listing 4.26, which contains an operation performing a **HZH** transformation sequence (Fig. 4.8).

```
operation MeasureHZH() : Result {
    use qubit = Qubit();
    H(qubit);
```

Fig. 4.8 HZH operation
sequence

```
    Z(qubit);
    H(qubit);
    let result = M(qubit);
    return result;
}
```
Listing 4.26 Q# operation verifying the relationship between the **Z** and **H** gates.

When invoked 4096 times using the standard sampling code this chapter has been using, the output shows that the transformations really result in an **X** bit-flip.

```
Measuring HZH
Measurement results:
Result 0: 0
Result 1: 4096
```
Listing 4.27 Expected output of the operation from Listing 4.26.

In addition to having their own standalone gate representations in Q#, the core library of Q# also provides a generic ApplyP (pauli : Pauli, target : Qubit) operation in the Microsoft.Quantum.Canon namespace. This is particularly useful in situations which require a conditional choice between selecting one of the Pauli gates, as it allows applying any Pauli gate in a parameterized fashion, such as shown in Listing 4.28.

```
// the following two statements are equivalent
Z(qubit);
ApplyP(PauliZ, qubit);
```
Listing 4.28 Example of a generic Pauli Q# operation.

For example if we want to conditionally setup a $|1\rangle$ state based on a boolean flag, instead of wrapping the **X** into a conditional clause like shown in the snippet below

```
if (setupOneState) {
    X(qubit);
}
```
Listing 4.29 An intuitive way of creating a $|1\rangle$ state with Q#.

it is possible to instead invoke the generic ApplyP (pauli : Pauli, target : Qubit) operation and use that boolean flag conditionally switch between PauliI and PauliX. This may be considered more concise and readable to some developers, as it reduces nesting of the code.

```
ApplyP(setupOneState ? PauliX | PauliI, qubit);
```
Listing 4.30 Creating a $|1\rangle$ state with Q# using the built-in operation to apply Pauli.

4.5   Rotation Gates

In Chap. 2 we mentioned that after a unitary transformation the transformed quantum
state vector remains normalized, since unitary transformations preserve the inner
product. As such, every unitary transformation can be viewed as a vector rotation.

In quantum computing, there is nevertheless a specialized set of gates explicitly
referred to as *rotation gates*. These three rotation gates, called $\mathbf{R}_x^{\varphi}$, $\mathbf{R}_y^{\varphi}$ and $\mathbf{R}_z^{\varphi}$, are
generalizations of the the the Pauli $\mathbf{X}$, $\mathbf{Y}$ and $\mathbf{Z}$ gates. The names of these gates come
form the fact that they allow a rotation around x-, y- and z-axes around the Bloch
sphere by an arbitrary angle.

4.5.1 $\mathbf{R}_z^{\varphi}$ Gate

Let us first consider $\mathbf{R}_z^{\varphi}$, which is a generalization of the $\mathbf{Z}$ gate. The gate allows rotating around the z-axis by a value of φ radians, changing the qubit phase accordingly
(Fig. 4.9).

The matrix representation of the gate is the following:

$$\mathbf{R}_z^{\varphi} = \begin{bmatrix} e^{-i\frac{\varphi}{2}} & 0 \\ 0 & e^{i\frac{\varphi}{2}} \end{bmatrix} \tag{4.44}$$

There is also an alternative form of expressing $\mathbf{R}_z^{\varphi}$, where the matrix from Eq. 4.44
is multiplied by a global phase $e^{i\frac{\varphi}{2}}$. Global phase is in principle unobservable and
hence is often omitted for simplicity [6].

$$\mathbf{R}z^{\varphi} = e^{i\frac{\varphi}{2}} \begin{bmatrix} e^{-i\frac{\varphi}{2}} & 0 \\ 0 & e^{i\frac{\varphi}{2}} \end{bmatrix} = \begin{bmatrix} 1 & 0 \\ 0 & e^{i\varphi} \end{bmatrix} \tag{4.45}$$

Due to this simplicity, we shall use the form from Eq. 4.45 in this book when
referring to $\mathbf{R}z^{\varphi}$.

Q# provides implementation of both variants. The variant from Eq. 4.45 is implemented as operation R1 (theta : Double, qubit : Qubit), while the one
from Listing 4.44 is built into the Q# standard library as Rz (theta : Double,
qubit : Qubit), both in the Microsoft.Quantum.Intrinsic namespace.

When inspecting the matrix composition of the gate Listing 4.45, we can quickly
realize two things. Because the following statements are true

$$e^0 = 1 \qquad e^{i\pi} = -1 \tag{4.46}$$

when $\varphi = 0$, $\mathbf{R}_z^{\varphi}$ becomes the identity gate, while when $\varphi = \pi$ it becomes the $\mathbf{Z}$ gate.

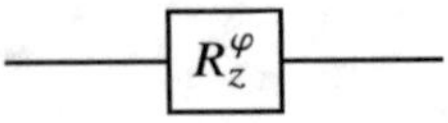

Fig. 4.9 $\mathbf{R}_z^{\varphi}$ rotation gate

$$-\boxed{R_z^{\varphi}}-$$

$$\mathbf{R_z^0} = \mathbf{I} \tag{4.47}$$
$$\mathbf{R_z^{\pi}} = \mathbf{Z}$$

In other words, both $\mathbf{I}$ and $\mathbf{Z}$ are special cases of the general z-axis rotation gate. There are two other commonly used special cases for $\mathbf{R_z^{\varphi}}$—when $\varphi = \frac{\pi}{2}$ and when $\varphi = \frac{\pi}{4}$.

4.5.2 S Gate

The first one is referred to as the **S** gate:

Fig. 4.10 S gate

Because of another special behavior of e, namely:

$$e^{\frac{\pi i}{2}} = i \tag{4.48}$$

substituting $\varphi = \frac{\pi}{2}$ into the $\mathbf{R_z^{\varphi}}$ gate from (4.45) produces the following matrix representation for **S**:

$$\mathbf{S} = \begin{bmatrix} 1 & 0 \\ 0 & i \end{bmatrix} \tag{4.49}$$

Since **S** is a rotation by $\frac{\pi}{2}$, it is performing half of a rotation of what the **Z** gate would do. This leads to an interesting relationship, where **S** applied twice in a row, produces the **Z** gate. A slightly less intuitive way of thinking about it is that **S** is a square root of **Z**.

$$\mathbf{SS} = \mathbf{S}^2 = \mathbf{Z} \tag{4.50}$$
$$\mathbf{S} = \sqrt{\mathbf{Z}}$$

Using the same diagnostics features of the Q# simulator that were introduced when discussing the **Z** gate, it is possible to see up close the effects of the transformation. Since we already know that the phase transformation has no effect in the $|0\rangle$ state, it is enough to only consider the $|1\rangle$ state as the input. The following Q# snippet can be used for that (Fig. 4.10).

```
operation TestSAgainst1() : Unit {
    use qubit = Qubit();
    X(qubit);
    S(qubit);
    DumpMachine();
    Reset(qubit);
}
```

Listing 4.31 Q# code testing the behavior of the S gate.

The output of the operation should resemble the output from Listing 4.21, with the only difference being visible in the numeric value of the phase, which should be equal to $\frac{\pi}{2}$, instead of π in the case of the **Z** transformation.

```
Testing S against |1>
# wave function for qubits with ids: 0
|0>:0.0000 + 0.0000i ==                    [0.0000]
|1>:0.0000 + 1.0000i == ********** [1.0000]           ^      [1.57080 rad]
```
Listing 4.32 Sample output from the code from Listing 4.31.

Because of the **S** and **Z** relationship shown in Eq. 4.50, it is clear that **S** gate is not its own inverse, and therefore applying it twice in a row will not return the quantum state vector to the original state. The inverse of **S** is, as we defined in Eq. 2.34, $\mathbf{S}^\dagger$, a complex conjugate transpose of **S**.

$$\mathbf{S}^\dagger = \begin{bmatrix} 1 & 0 \\ 0 & -i \end{bmatrix} \tag{4.51}$$

We can calculate algebraically that multiplying the two matrices together

$$\mathbf{SS}^\dagger = I \tag{4.52}$$

produces an identity transformation—after all $-i \cdot i = 1$. Naturally, it is yet again tempting to see it in action in a Q# program. This can act as kind of a verification test to ensure that the applied reasoning and subsequent calculations are indeed correct (Fig. 4.11).

In Q# the built-in `Adjoint` functor, which we already briefly covered in Sect. 3.2, allows invocation of the Hermitian conjugate of any operation that supports them— typically one that was declared with the *adjoint* characteristic. As such, the code from Listing 4.31, updated to include $\mathbf{S}^\dagger$, looks like this:

```
operation TestSReversibility() : Unit {
    use qubit = Qubit();
    X(qubit);
    S(qubit);
    Adjoint S(qubit);
    DumpMachine();
    Reset(qubit);
}
```
Listing 4.33 Q# code testing the reversibility of the S gate.

The circuit representation for this code example is:

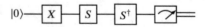

Fig. 4.11 Verifying the reversibility of the S gate

The output of such operation shows clearly that applying $\mathbf{S}^\dagger$ after $\mathbf{S}$ returned the qubit back to the initial state—which in this particular case is state $|1\rangle$, with the phase equal to zero.

```
Testing S reversibility
# wave function for qubits with ids: 0
|0>:0.0000 + 0.0000i ==                    [0.0000]
|1>:1.0000 + 0.0000i == ********** [1.0000]     --- [0.0000 rad]
```

Listing 4.34 Sample output from the code from Listing 4.33.

4.5.3 T Gate

Another special variant of the $\mathbf{R}_z^\varphi$ rotation gate is known as the **T** gate, which applies a z-axis rotation of the qubit by $\frac{\pi}{4}$ (Fig. 4.12).

Fig. 4.12 T gate

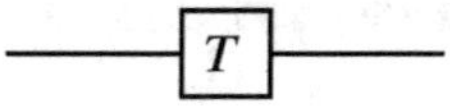

And the matrix form of it looks as follows:

$$\mathbf{T} = \begin{bmatrix} 1 & 0 \\ 0 & e^{\frac{\pi}{4}} \end{bmatrix} \tag{4.53}$$

Just like **S** applies half of the rotation of the **Z** gate, **T** applies half of the rotation of the **S**. Because of that we can say that the following statements are all true:

$$\mathbf{TT} = \mathbf{T}^2 = \mathbf{S} \tag{4.54}$$
$$\mathbf{TTTT} = \mathbf{T}^4 = \mathbf{S}^2 = \mathbf{Z}$$
$$\mathbf{T} = \sqrt{\mathbf{S}} = \sqrt[4]{\mathbf{Z}}$$

To double check if this reasoning is really correct, we shall go back to Q# now and try to confirm that applying **T** gate on qubit four times in a row, really produces the same effects as $\mathbf{S}^2$ from Listing 4.32 or **Z** from Listing 4.23. The relevant Q# code is shown below.

```
operation TestTZRelationship() : Unit {
    use qubit = Qubit();
    X(qubit);
    T(qubit);
    T(qubit);
    T(qubit);
    T(qubit);
    DumpMachine();
    Reset(qubit);
}
```

Listing 4.35 Q# code testing the relationship between the **T** and **Z** gates.

This produces a state output that indeed is identical to the state output of the expected rotation of **Z**, confirming empirically that they are really equal to each other.

One final thing worth mentioning at this point is that the code in listing above is not particularly elegant—the application of **T** four times in a row seems rather clumsy, if not to say gruesome. Thankfully Q# has relevant tools to help with that, namely it provides a mechanism for automatically creating composite operations based on powers of an individual operation.

In this particular case, we can leverage the function `OperationPow<'T>` `(op : ('T => Unit), power : Int)` from the Microsoft. `Quantum.Canon` namespace, and let it create a compound operation representing T^4. Afterwards, we only need to apply the resulting operation to the qubit once. Such technique is called *operation exponantiation* and is shown in Listing 4.36 and we shall make further use of it in Chap. 7.

```
let t4 = OperationPow(T, 4);
t4(qubit);
```

Listing 4.36 Using the built-in feature of Q# standard library to create operation exponantiation.

4.5.4 R_x^φ and R_y^φ Gates

The two rotation gates $\mathbf{R}_x^\varphi$ and $\mathbf{R}_y^\varphi$ are generalizations of the the **X** and **Y** gates, performing rotations around x- and y-axes respectively. This is of course analogous to $\mathbf{R}_z^\varphi$ and its relationship to **Z** (Fig. 4.13).

$\mathbf{R}_x^\varphi$ and $\mathbf{R}_y^\varphi$ are described by the matrices below

$$\mathbf{R}_x^\varphi = \begin{bmatrix} cos(\frac{\varphi}{2}) & -sin(\frac{\varphi}{2})i \\ -sin(\frac{\varphi}{2})i & cos(\frac{\varphi}{2}) \end{bmatrix} \tag{4.55}$$

$$\mathbf{R}_y^\varphi = \begin{bmatrix} cos(\frac{\varphi}{2}) & -sin(\frac{\varphi}{2}) \\ sin(\frac{\varphi}{2})i & cos(\frac{\varphi}{2}) \end{bmatrix} \tag{4.56}$$

For the purpose of fine grained control over the qubit rotations, Q# provides implementations of all three rotation gates—$\mathbf{R}_z^\varphi$ (which we already covered, along with its R1 cousin), $\mathbf{R}_x^\varphi$ and $\mathbf{R}_y^\varphi$. They are part of the Q# standard library under the `Microsoft.Quantum.Intrinsic` namespace, and all have very similar signatures. The desired rotation angle is specified in radians (Fig. 4.14).

```
operation Rx(theta : Double, qubit : Qubit) : Unit is Adj + Ctl
operation Ry(theta : Double, qubit : Qubit) : Unit is Adj + Ctl
operation Rz(theta : Double, qubit : Qubit) : Unit is Adj + Ctl
```

Listing 4.37 Signatures of the built-in Q# rotation gates about all three axes.

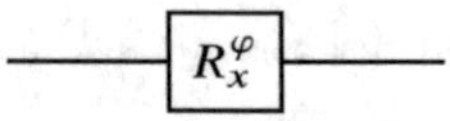

Fig. 4.13 R_x^φ rotation gate with an angle of φ

Fig. 4.14 R_y^φ rotation gate with an angle of φ

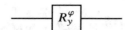

There is an additional general rotation operation that can be used in place of the three specialized variants—in that case the axis becomes an argument that is passed to the operation as well. This conceptually similar to the generalized measurement we looked at in Listing 4.4, where the Pauli basis was a parameter—here it is possible to dynamically choose the rotation axis using the Pauli keyword literal.

```
operation R(pauli : Pauli, theta : Double, qubit : Qubit) : Unit
    is Adj + Ctl
```

Listing 4.38 General rotation operation in Q#.

When the R operation from Eq. 4.38 is invoked with PauliI provided as parameter, the QDK framework only applies a global phase to the qubit. This can be useful to compensate for global phase changes. For example the aforementioned R1 operation, defined in Eq. 4.45, is internally implemented by first performing a z-axis rotation and then uses R operation with PauliI and a negative rotation angle to undo the global phase change that was applied during the R_z^φ.

4.6 Multi Qubit Gates

In Sect. 4.2, we defined quantum gates as $2^n \times 2^n$ sized unitary matrices, where n stands for the number of qubits the gate acts on. So far we only dealt with single qubit gates, though there are not many algorithms that can be built out of single qubit gates only. To address that, in this section we shall introduce some gates that can span multiple qubits—they will quickly become an invaluable part of our quantum toolbox.

The tensor product formalism for describing composite quantum systems, which we first introduced in Sect. 2.6, tells us that for two qubits, the quantum state is described by a four-dimensional abstract complex vector. In general, the overall state of this two qubit system is expressed by the tensor product of those two qubits $|\psi_1\rangle$ and $|\psi_2\rangle$

$$|\psi_1\rangle = a_1 |0\rangle + b_1 |1\rangle \qquad |\psi_2\rangle = a_2 |0\rangle + b_2 |1\rangle$$
$$|\psi_1\rangle \otimes |\psi_2\rangle = a_1a_2 |0\rangle |0\rangle + a_1b_2 |0\rangle |1\rangle + b_1a_2 |1\rangle |0\rangle + b_1b_2 |1\rangle |1\rangle \qquad (4.57)$$

Thus, applying the tensor product between the computational basis vectors, we get four dimensional standard basis vectors for two qubit systems

$$|00\rangle = \begin{bmatrix} 1 \\ 0 \\ 0 \\ 0 \end{bmatrix} \qquad |01\rangle = \begin{bmatrix} 0 \\ 1 \\ 0 \\ 0 \end{bmatrix} \qquad |10\rangle = \begin{bmatrix} 0 \\ 0 \\ 1 \\ 0 \end{bmatrix} \qquad |11\rangle = \begin{bmatrix} 0 \\ 0 \\ 0 \\ 1 \end{bmatrix} \qquad (4.58)$$

Instead of using binary notation, it is sometimes useful to label the basis state kets using unsigned integers which the binary representations of each state represent. In such notation, the two qubit standard basis looks as follows:

$$\{|00\rangle , |01\rangle , |10\rangle , |11\rangle\} = \{|0\rangle , |1\rangle , |2\rangle , |3\rangle\} \tag{4.59}$$

This naturally can be extrapolated further—a standard basis for three qubit systems would be:

$$\{|000\rangle , |001\rangle , |010\rangle , |011\rangle , |100\rangle , |101\rangle , |110\rangle , |111\rangle\} =$$
$$\{|0\rangle , |1\rangle , |2\rangle , |3\rangle , |4\rangle , |5\rangle , |6\rangle , |7\rangle\} \tag{4.60}$$

Additionally, individual single qubit unitary transformations can be combined into a single multi-qubit transformation using tensor products between those gates. For example consider the circuit below, where an **X** gate is applied to the first qubit, while the **Z** gate is applied to the second one (Fig. 4.15).

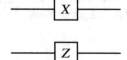

Fig. 4.15 Two qubit circuit with **X** and **Z** transformations

We can define here a new two qubit unitary transformation such that:

$$\mathbf{U_{xz}} = \mathbf{X} \otimes \mathbf{Z} = (\mathbf{X} \otimes \mathbf{I})(\mathbf{I} \otimes \mathbf{Z}) \tag{4.61}$$

Such combined new transformation can now be depicted on a circuit as one that really spans across two qubits (Fig. 4.16).

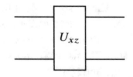

Fig. 4.16 Two qubit circuit with a single transformation encompassing **X** and **Z** transformations

4.6.1   Controlled Gates

While we can express individual transformations as a composite tensor-based transformation, we cannot always decompose a multi-qubit transformation into a set of single qubit ones. This class of transformations are called controlled transformations—where the state of one of the qubits is conditionally changed based on the state of other qubits making up the composite system. In particular, if the qubits are not in a definite basis state but rather in a superposition state, these transformations create *entanglement*.

Entanglement between the qubits is a critical concept in quantum information theory, however, for the time being, we shall only consider the two computational basis states $|0\rangle$ and $|1\rangle$, which will allow us to slowly, but diligently, uncover the underlying theoretical scheme behind the multi-qubit gates. We shall then dedicate the entire Chap. 5 to entanglement.

An intuitive way of thinking about controlled gates is that they are made up of two separate transformations. First, there is the identity gate **I**, making up the top left corner of the matrix representing the entire two-qubit gate, and acting on the first qubit, commonly referred to as *control* qubit. This implies that the control qubit is unchanged by the controlled gate. Second is a single qubit gate, conditionally acting on the qubit aptly referred to as the *target*, and it can be found in the bottom right corner of the two-qubit matrix. The reason why we say that the target state transformation is applied conditionally, is because it will only be invoked if the control qubit is in the definite basis state $|1\rangle$.

Let us now define a generic single qubit gate **U**:

$$\mathbf{U} = \begin{bmatrix} U_{11} & U_{12} \\ U_{21} & U_{22} \end{bmatrix} \tag{4.62}$$

If we follow the generalization path for two-qubit controlled gates we just discussed, it is now possible to define a two-qubit controlled-**U**, or **CU**, such that:

$$\mathbf{CU} = \begin{bmatrix} 1 & 0 & 0 & 0 \\ 0 & 1 & 0 & 0 \\ 0 & 0 & U_{11} & U_{12} \\ 0 & 0 & U_{21} & U_{22} \end{bmatrix} \tag{4.63}$$

We can now classify the effects of this CU gate in the following way:

$$\mathbf{CU}\,|00\rangle = |00\rangle \qquad \mathbf{CU}\,|01\rangle = |01\rangle \tag{4.64}$$
$$\mathbf{CU}\,|10\rangle = |1\rangle \otimes \mathbf{U}\,|0\rangle \qquad \mathbf{CU}\,|11\rangle = |1\rangle \otimes \mathbf{U}\,|1\rangle$$

When the control is in state $|0\rangle$, the target state is not transformed, and thus the entire gate has no effects on the system. On the other hand, when the control is in state $|1\rangle$, control remains unchanged, but target gets transformed by **U**.

The common way of depicting controlled gates on quantum circuits is to use a thick dot over the control qubit, then have a vertical line connecting the control qubit to a target qubit, and then the regular single gate symbol over the target qubit. An example for **CU** is shown in Fig. 4.17.

Q# allows for automatic generation of controlled operations from regular operation by declaring the *controlled* specialization on an them. This is a concept that we already introduced in Sect. 3.2, where we also showed how to achieve this by employing one of three possible syntax constructs. The annotation based approach,

Fig. 4.17 A sample
controlled **U** gate

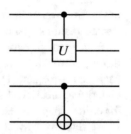

Fig. 4.18 CNOT gate

typically referred to as specifying the *operation characteristics*, is the most compact
and arguably most readable one, and is the one we will use throughout this book.

```
operation QuantumTransformation(q : Qubit) : Unit is Ctl {
    // operation body
}
```

Listing 4.39 Declaring the controlled specialization on a Q# operation using an operation
characteristic.

Such controlled operations can then be invoked using the `Controlled` functor.

```
use (control, target) = (Qubit(), Qubit());
Controlled QuantumTransformation([control], target);
```

Listing 4.40 Example of invoking the controlled operation via the functor.

4.6.2   CNOT *Gate*

The most fundamental and most commonly used two-qubit gate is the **CNOT** gate. It
applies the controlled version of the **X** transformation and therefore in some literature
it is also referred to as a **CX** gate. **CNOT** is mathematically expressed using the matrix
shown in Eq. 4.65.

$$
\mathbf{CNOT} = \begin{bmatrix} 1 & 0 & 0 & 0 \\ 0 & 1 & 0 & 0 \\ 0 & 0 & 0 & 1 \\ 0 & 0 & 1 & 0 \end{bmatrix} \tag{4.65}
$$

Based on Eq. 4.63 and the fact that single qubit **X** gate is a bit-flip gate, we already
know that upon application of **CNOT**, the quantum state of the control qubit will
not change, while the target qubit, depending on the state of the control, may be
conditionally transformed into its own opposite (Fig. 4.18).

As long as we only consider the two computational basis states, a good analogy
for the logic executed by **CNOT** is the XOR, exclusive OR, operation from classical
computing. For two classic bit inputs XOR returns 0 when the two inputs are both 0
or both 1, and 1 when the two input values differ—one is a 0 and the other a 1. This

can be written as:

$$0 \oplus 0 = 0 \quad 0 \oplus 1 = 1 \quad 1 \oplus 0 = 1 \quad 1 \oplus 1 = 0 \tag{4.66}$$

If we now leave aside the quantum mechanical concept of a possible superposition of either control and target, and the complexities arising from that, and only consider the two basis states $|0\rangle$ and $|1\rangle$,[9] we can easily verify that the effect of the **CNOT** gate actually aligns perfectly with classical XOR.

$$
\begin{aligned}
|00\rangle &\rightarrow |0\rangle \otimes |0 \oplus 0\rangle \rightarrow |00\rangle \\
|01\rangle &\rightarrow |0\rangle \otimes |0 \oplus 1\rangle \rightarrow |01\rangle \\
|10\rangle &\rightarrow |1\rangle \otimes |1 \oplus 0\rangle \rightarrow |11\rangle \\
|11\rangle &\rightarrow |1\rangle \otimes |1 \oplus 1\rangle \rightarrow |10\rangle
\end{aligned}
\tag{4.67}
$$

Such division of roles between a control qubit and a target qubit is not definite. The model discussed so far, while helpful for high-level understanding, should not be taken at face value, because it only applies to the standard basis. To be more precise, so far we only considered the **CNOT** behavior in the standard computational basis $\{|0\rangle, |1\rangle\}$. However, it is equally valid to apply the gate in a different basis, say, the X-basis $\{|+\rangle, |-\rangle\}$. In the X-basis, the notions of control and target qubits are effectively flipped, as it is the second qubit—the one we so far referred to as the "target"—that remains unchanged, and the first one—the "control"—changes its state conditionally. The **CNOT** state transformations for the basis states of the X-basis looks the following:

$$
\begin{aligned}
|++\rangle &\rightarrow |++\rangle \\
|+-\rangle &\rightarrow |--\rangle \\
|-+\rangle &\rightarrow |-+\rangle \\
|--\rangle &\rightarrow |+-\rangle
\end{aligned}
\tag{4.68}
$$

Because **X** matrix is Hermitian, the **CNOT** gate is also Hermitian, which means it is both unitary and also its own inverse. As a consequence, it can be applied twice in a row to reverse its result. In other words, given an initial quantum state $|\psi\rangle$, the following relation holds:

$$\mathbf{CNOT}(\mathbf{CNOT}\,|\psi\rangle) = |\psi\rangle \tag{4.69}$$

This can proven in several simple steps. Suppose we start with a two-qubit composite quantum state $|\psi_1\rangle$:

$$|\psi_1\rangle = |\varphi_1\rangle \otimes |\varphi_2\rangle \tag{4.70}$$

---

[9] When considering the basis states only, quantum computer behaves like a classical computer.

Let us now assume that the control qubit is the one in the state φ_1 and the target qubit is in the state φ_2. Applying **CNOT** gate to $|\psi_1\rangle$ will transform the system state to $|\psi_2\rangle$:

$$|\psi_2\rangle = \textbf{CNOT}\,|\psi_1\rangle = |\varphi_1\rangle \otimes |\varphi_1 \oplus \varphi_2\rangle \tag{4.71}$$

Applying **CNOT** transformation to the resulting state $|\psi_2\rangle$ again, will further transform it to $|\psi_3\rangle$:

$$|\psi_3\rangle = \textbf{CNOT}\,|\psi_2\rangle = |\varphi_1\rangle \otimes |\varphi_1 \oplus (\varphi_1 \oplus \varphi_2)\rangle \tag{4.72}$$

Since the XOR logic is commutative and associative, we can actually replace $\varphi_1 \oplus \varphi_1$ with 0—recall that XOR returns 0 when both inputs are the same—which leads to the representation $|\psi_3\rangle$:

$$|\psi_3\rangle = |\varphi_1\rangle \otimes |0 \oplus \varphi_2\rangle = |\varphi_1\rangle \otimes |\varphi_2\rangle \tag{4.73}$$

and that is really equivalent to the initial state $|\psi_1\rangle$ from (4.70).

Q# naturally ships with a built-in implementation of the **CNOT** gate, but before we explore that, let us use the Q# 's `Controlled` functor to manually construct a **CNOT** from the single qubit **X** gate. We would like to test the outcome of such gate for input being four possible basis states for a two-qubit system—$|00\rangle$, $|01\rangle$, $|10\rangle$ and $|11\rangle$. The example operation is shown in the Listing 4.41.

```
operation ManualCNOT(initState : Bool[]) : Unit {
    use (control, target) = (Qubit(), Qubit());
    ApplyP(initState[0] ? PauliX | PauliI, control);
    ApplyP(initState[1] ? PauliX | PauliI, target);

    Controlled X([control], target);
    let result = MultiM([control, target]);
    PrintResult(initState, result);
}
```
Listing 4.41 Constructing a **CNOT** from a controlled functor and an **X** gate

The code listing takes advantage of several things worth explaining. First of all, since we would like to test various combinations of the initial system state, the operation allows parameterized input that will determine the states of control and target qubit. This can easily be done with boolean flags, which could be set to true in case a given qubit needs to be in state $|1\rangle$.

This conditional transformation into the basis state $|1\rangle$, where necessary, is achieved by using the code discussed previously in Listing 4.30—namely by invoking the `ApplyP (pauli : Pauli, target : Qubit)` operation with the appropriate Pauli keyword literal. The measurement, in order to avoid the verbosity related to individual qubit measurements is done using the `MultiM (targets : Qubit[])` operation, which was introduced previously in Sect. 4.3.

Finally, a helper function from Listing 4.42 to print out the results is used. It is technically not necessary but is a nice way to help keep the operation itself focused on the important aspects of the quantum logic only. We shall reuse this function in other parts of this chapter as well—it converts the `Result` and `Bool` arrays into user friendly console output messages.

```
function PrintResult(initialState : Bool[], measuredState :
    Result[]) : Unit {
    mutable input = "";
    mutable output = "";
    for i in 0..Length(initialState)-1 {
        set input += (initialState[i] ? "1" | "0");
        set output += (measuredState[i] == One ? "1" | "0");
    }

    Message($"|{input}> ==> |{output}>");
}
```

Listing 4.42 Helper function to print out testing results.

We would want to invoke the operation we just set up in Listing 4.41 four times—providing $|00\rangle$, $|01\rangle$, $|10\rangle$ and $|11\rangle$ as input states. The output should be exactly the same as the expected state transitions that we defined as part of Listing 4.67.

```
@EntryPoint()
operation Main() : Unit {
    ManualCNOT([false, false]); // |00>
    ManualCNOT([false, true]);  // |01>
    ManualCNOT([true, false]);  // |10>
    ManualCNOT([true, true]);   // |11>
}
```

Listing 4.43 Orchestration code verifying the code from Listing 4.41.

```
|00> ==> |00>
|01> ==> |01>
|10> ==> |11>
|11> ==> |10>
```

Listing 4.44 Expected output from the code in Listing 4.43.

Code from Listing 4.41 can also be easily modified to replace the manual construction of the **CNOT** gate from the `Controlled` functor and an **X** gate with the built-in `CNOT (control : Qubit, target : Qubit)` operation. The updated code is shown in Listing 4.45.

```
operation LibraryCNOT(initState : Bool[]) : Unit {
    use (control, target) = (Qubit(), Qubit());
    ApplyP(initState[0] ? PauliX | PauliI, control);
    ApplyP(initState[1] ? PauliX | PauliI, target);

    CNOT(control, target);
    let result = MultiM([control, target]);
    PrintResult(initState, result);
}
```

Listing 4.45 Orchestrating the Q# built-in **CNOT** implementation.

4.6.3 SWAP *Gate*

Another commonly used two-qubit gate is the **SWAP** gate. As the name implies, it can be applied in order to swap the quantum states of two qubits. This is a feature that is commonly relied upon in various quantum algorithms, especially given that an unknown quantum state cannot be copied. We will utilize the SWAP gate in later chapters (Fig. 4.19).

SWAP is described by the following matrix:

$$\mathbf{SWAP} = \begin{bmatrix} 1 & 0 & 0 & 0 \\ 0 & 0 & 1 & 0 \\ 0 & 1 & 0 & 0 \\ 0 & 0 & 0 & 1 \end{bmatrix} \tag{4.74}$$

The effect of **SWAP** is similar to the effect of the single qubit **X** gate. Recall that the **X** gate effectively swaps the probability amplitudes of $|0\rangle$ and $|1\rangle$:

$$\mathbf{X}(a\,|0\rangle + b\,|1\rangle) = b\,|0\rangle + a\,|1\rangle \tag{4.75}$$

For a two qubit composite quantum system, in the computational basis, the **SWAP** gate actually swaps the probability amplitudes of the coefficients of $|01\rangle$ and $|10\rangle$. Such outcome is a consequence of applying the tensor product between the subsystems in the reverse order.

$$\mathbf{SWAP}(a_1 a_2\,|00\rangle + a_1 b_2\,|01\rangle + b_1 a_2\,|10\rangle + b_1 b_2\,|11\rangle) =$$
$$a_2 a_1\,|00\rangle + a_2 b_1\,|01\rangle + b_2 a_1\,|10\rangle + b_2 b_1\,|11\rangle \tag{4.76}$$

As a result, **SWAP** state transitions for the computational basis states $|0\rangle$ and $|1\rangle$ can be summarized as:

$$
\begin{array}{ll}
|00\rangle \rightarrow |00\rangle & |01\rangle \rightarrow |10\rangle \\
|10\rangle \rightarrow |01\rangle & |11\rangle \rightarrow |11\rangle
\end{array} \tag{4.77}
$$

Fig. 4.19 SWAP gate

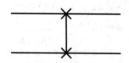

Fig. 4.20 SWAP gate
reconstructed from a
sequence of three **CNOT**
gates

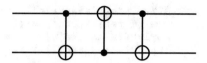

The **SWAP** gate can be accurately reconstructed using a series of three **CNOT** transformations, applied in the order shown on the circuit in Fig. 4.20.

While such composition may not appear very intuitive at first, it can be proven, similarly to our earlier **CNOT** reversibility proof, relatively easily. Let us imagine having an initial state $|\psi_1\rangle$ of the composite two-qubit system:

$$|\psi_1\rangle = |\varphi_1\rangle \otimes |\varphi_2\rangle \tag{4.78}$$

After applying the first **CNOT** we end up with the state $|\psi_2\rangle$:

$$|\psi_2\rangle = \mathbf{CNOT}\,|\psi_1\rangle = |\varphi_1\rangle \otimes |\varphi_1 \oplus \varphi_2\rangle \tag{4.79}$$

Following with the second **CNOT**, this time with reversal of control qubit, we end up with $|\psi_3\rangle$. Notice that the reversal causes the state that was already described by XOR to stay intact.

$$|\psi_3\rangle = \mathbf{CNOT}\,|\psi_2\rangle = |\varphi_1 \oplus (\varphi_1 \oplus \varphi_2)\rangle \otimes |\varphi_1 \oplus \varphi_2\rangle \tag{4.80}$$

Due to associativity of XOR, $|\psi_3\rangle$ can then be reduced to:

$$|\psi_3\rangle = |\varphi_2\rangle \otimes |\varphi_1 \oplus \varphi_2\rangle \tag{4.81}$$

Finally, applying the final **CNOT** (reversing the control qubit again) leads us to the final state $|\psi_4\rangle$:

$$|\psi_4\rangle = \mathbf{CNOT}\,|\psi_3\rangle = |\varphi_2\rangle \otimes |(\varphi_1 \oplus \varphi_2) \oplus \varphi_2\rangle \tag{4.82}$$

And just like before, this can be reduced to:

$$|\psi_4\rangle = |\varphi_2\rangle \otimes |\varphi_1\rangle \tag{4.83}$$

State $|\psi_4\rangle$ is the same tensor product as in $|\psi_1\rangle$, except applied in reverse order, which is precisely the definition of **SWAP**.

$$\mathbf{SWAP}(|\varphi_1\rangle \otimes |\varphi_2\rangle) = |\varphi_2\rangle \otimes |\varphi_1\rangle \tag{4.84}$$

SWAP gate is implemented in Q# in the form of a `SWAP` (`qubit1 : Qubit`, `qubit2 : Qubit`) operation in the `Microsoft.Quantum.Intrinsic` namespace. We can test it using by making only a small modification to the code from Listing 4.45—by replacing the **CNOT** call with **SWAP**:

```
operation LibrarySWAP(initState : Bool[]) : Unit {
    use (first, second) = (Qubit(), Qubit());
    ApplyP(initState[0] ? PauliX | PauliI, first);
    ApplyP(initState[1] ? PauliX | PauliI, second);

    SWAP(first, second);
    let result = MultiM([first, second]);
    PrintResult(initState, result);
}
```
Listing 4.46 Usage of the built-in **SWAP** gate.

The output of this code, when ran for four different input states: $|00\rangle$, $|01\rangle$, $|10\rangle$ and $|11\rangle$ would be:

```
|00> ==> |00>
|01> ==> |10>
|10> ==> |01>
|11> ==> |11>
```
Listing 4.47 Output of the code from Listing 4.46.

We are also going to implement **SWAP** in Q# manually, however, we will not do it by simply calling the built-in **CNOT** operation three times in a row—that would be too simple. Instead, we will take this opportunity to explore the *conjugation* feature of Q#, which we initially introduced in Sect. 3.2. Given that **CNOT** is Hermitian we can take advantage of the Q# conjugation syntax to reconstruct a **SWAP** gate by writing:

```
within {
    CNOT(first, second);
}
apply {
    CNOT(second, first);
}
```
Listing 4.48 **SWAP** constructed from three **CNOT** gates using the conjugation feature of Q#.

The compiler here would generate the $U^{\dagger}VU$ pattern, which due to the fact that **CNOT** is Hermitian will result in a de facto application of three **CNOT** gates in a row, exactly how we pictured it on Fig. 4.20. The rest of the code would be identical to Listing 4.46 so we shall not repeat it here again.

4.6.4 CZ Gate

Similarly to **CNOT** being a controlled variant of the single qubit **X** gate, **CZ** is a gate that allows applying, in a controlled fashion, the **Z** transform (Fig. 4.21).

As we already learnt, **Z** gate performs a rotation around the z-axis by π radians, resulting in the so-called phase change (sign flip) of the qubit. Just like the entire category of controlled transformations, **CZ** applies the **Z** phase flip onto the target when the control qubit is in a basis state $|1\rangle$. The single qubit **Z** transformation, in the computational basis, only acts on the amplitude associated with $|1\rangle$ and, similarly, in the **CZ**, the only affected state is $|11\rangle$.

CZ is described by the following matrix:

$$\mathbf{CZ} = \begin{bmatrix} 1 & 0 & 0 & 0 \\ 0 & 1 & 0 & 0 \\ 0 & 0 & 1 & 0 \\ 0 & 0 & 0 & -1 \end{bmatrix} \tag{4.85}$$

It is easy to notice that, as it is the case with other controlled gates, **CZ** consists of the **I** matrix in the top left corner, and the relevant target transformation matrix, **Z** in this case, in the bottom right corner.

An interesting feature of **CZ** is that despite being a *controlled* gate, due to the fact that it always only impacts the $|11\rangle$ state, it actually does not matter which qubit is control and which one is target—the effect of the **CZ** transformation is always the same regardless.

Similarly to **Z** gate, which can be reconstructed from a combination of **X** and **H**, **CZ** can be expressed as a combination of **CNOT** and **H**

$$\mathbf{CZ} = (\mathbf{I} \otimes \mathbf{H})\mathbf{CNOT}(\mathbf{I} \otimes \mathbf{H}) \tag{4.86}$$

The general effect of the **CZ** on a two qubit quantum state, with a_1 and b_1 being the probability amplitudes for the first qubit, a_2 and b_2 for the second one, can be summarized as

$$\begin{aligned} \mathbf{CZ}(a_1a_2\,|00\rangle + a_1b_2\,|01\rangle + b_1a_2\,|10\rangle + b_1b_2\,|11\rangle) = \\ a_1a_2\,|00\rangle + a_1b_2\,|01\rangle + b_1a_2\,|10\rangle - b_1b_2\,|11\rangle) \end{aligned} \tag{4.87}$$

which of course is the sign flip of the amplitude related to $|11\rangle$. Consequently, **CZ** state transitions for the computational basis states $|0\rangle$ and $|1\rangle$ can be written as:

$$\begin{aligned} |00\rangle &\to |00\rangle & |01\rangle &\to |01\rangle \\ |10\rangle &\to |10\rangle & |11\rangle &\to -|11\rangle \end{aligned} \tag{4.88}$$

Q# does not have a built-in **CZ** gate, but it can be easily constructed out of a combination of the built-in **Z** gate and the `Controlled` functor. The **CZ** gate can also be extend to span any number of qubits and would always only affect state $|11..1\rangle$.

Fig. 4.21 CZ gate

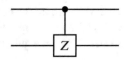

4.6.5 Toffoli Gate

Toffoli gate, also known as **CCNOT** gate, is a three-qubit doubly controlled gate. It is a generalization of **CNOT** over three-qubits and as such it behaves the same ways as **CNOT** does, that is, logically, it follows XOR rules. **CCNOT**, instead of having a single control qubit, has two control qubits, both of which must be in state $|1\rangle$ for the **X** transformation to be applied to the target. From that perspective, the gate only acts on states $|110\rangle$ and $|111\rangle$, leaving the others unchanged. We can generalize this as:

$$|x, y, z\rangle \rightarrow |x, y, (z \oplus xy)\rangle \tag{4.89}$$

Since the **CCNOT** gate operates on three qubits, which are described by an eight dimensional quantum state vector, the matrix used to express it is 8×8 in size:

$$\mathbf{CCNOT} = \begin{bmatrix} 1 & 0 & 0 & 0 & 0 & 0 & 0 & 0 \\ 0 & 1 & 0 & 0 & 0 & 0 & 0 & 0 \\ 0 & 0 & 1 & 0 & 0 & 0 & 0 & 0 \\ 0 & 0 & 0 & 1 & 0 & 0 & 0 & 0 \\ 0 & 0 & 0 & 0 & 1 & 0 & 0 & 0 \\ 0 & 0 & 0 & 0 & 0 & 1 & 0 & 0 \\ 0 & 0 & 0 & 0 & 0 & 0 & 0 & 1 \\ 0 & 0 & 0 & 0 & 0 & 0 & 1 & 0 \end{bmatrix} \tag{4.90}$$

All of the possible state transitions enacted by the **CCNOT** gate on the eight computational basis states are:

$$\begin{array}{ll} |000\rangle \rightarrow |000\rangle & |001\rangle \rightarrow |001\rangle \\ |010\rangle \rightarrow |010\rangle & |011\rangle \rightarrow |011\rangle \\ |100\rangle \rightarrow |100\rangle & |101\rangle \rightarrow |101\rangle \\ |110\rangle \rightarrow |111\rangle & |111\rangle \rightarrow |110\rangle \end{array} \tag{4.91}$$

In other words, the gate swaps the states $|110\rangle$ and $|111\rangle$ with each other (Fig. 4.22).

In Q#, just like the **CNOT**, **SWAP** and other basic gates, **CCNOT** is part of the `Microsoft.Quantum.Intrinsic` namespace and it happens to be the only three-qubit gate that is built-into the Q# standard library. However, other multi-qubit controlled gates can always be manually constructed using the aforementioned `Controlled` functor, which supports arbitrary sizes of controlled gates.

We can yet again reuse code from Listing 4.45, and test the Q# implementation of **CCNOT** with it. The only change required is to actually use the **CCNOT** in place of **CNOT**, and scale up the initial state setup, as well as the qubit allocation statements, from two to three qubits.

Fig. 4.22 CCNOT gate

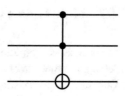

```
operation LibraryCCNOT(initState : Bool[]) : Unit {
    use (ctl1, ctl2, target) = (Qubit(), Qubit(), Qubit());
    ApplyP(initState[0] ? PauliX | PauliI, ctl1);
    ApplyP(initState[1] ? PauliX | PauliI, ctl2);
    ApplyP(initState[2] ? PauliX | PauliI, target);

    CCNOT(ctl1, ctl2, target);
    let result = MultiM([ctl1, ctl2, target]);
    PrintResult(initState, result);
}
```
Listing 4.49 Usage of the built-in **CCNOT** gate.

We will need to invoke this test operation eight times to capture all possible outcomes for the eight dimensional quantum state vector describing three qubits.

```
@EntryPoint()
operation Main() : Unit {
    LibraryCCNOT([false, false, false]); // |000>
    LibraryCCNOT([false, false, true]);  // |001>
    LibraryCCNOT([false, true, false]);  // |010>
    LibraryCCNOT([false, true, true]);   // |011>
    LibraryCCNOT([true, false, false]);  // |100>
    LibraryCCNOT([true, false, true]);   // |101>
    LibraryCCNOT([true, true, false]);   // |110>
    LibraryCCNOT([true, true, true]);    // |111>
}
```
Listing 4.50 Testing the built-in **CCNOT** gate.

The output of this code should be the same as the expectations we outlined in Eq. 4.91, namely we should only see the last two states $|110\rangle$ and $|111\rangle$ getting swapped with each other.

```
|000> ==> |000>
|001> ==> |001>
|010> ==> |010>
|011> ==> |011>
|100> ==> |100>
|101> ==> |101>
|110> ==> |111>
|111> ==> |110>
```
Listing 4.51 Expected output of code from Listing 4.50.

4.7   Gate Universality

In classical computation theory, we often speak of the *universal gates*—a minimum set of gates that can be arranged in a way that they can represent any logical algorithm. Conversely, any existing algorithm could theoretically be decomposed into such universal gate-based representation. Such (de)composition would be suboptimal from the performance standpoint, but it would nevertheless be functionally complete. In fact, a single classical gate, the NOT-AND gate, commonly referred to as the NAND gate, is such a universal gate. Using a combination of NAND gates alone, and no other helper gates, it is possible to express any boolean logic expression or function. In quantum computing, the concept of gate universality is a little more challenging.

As it turns out, quantum computing does not allow for a perfect set of universal gates. The main reason behind it, is that any arbitrary unitary transformation can become a gate that transforms the quantum state of a qubit. Consequently, this precludes the existence of a perfect set of universal gates. A helpful way of thinking about it, reported by Bernhardt [1], is to remember that due to the continuous nature of the single qubit rotations, there are infinitely many transformations that could be applied to a single qubit, yielding infinitely many potential circuits. On the other hand, a finite set of gates could only be combined in a finite possible ways.

Despite this rather sobering assessment of the state of affairs, Nielsen and Chuang [4] do identify a set of universal quantum gates. They do this by providing their own criteria for gate universality—introducing a quantum-specific definition of *gate universality*. They define a set of gates to be universal when any unitary operation could be "approximated to arbitrary accuracy by a quantum circuit involving only those gates". Those universal quantum gates are three single qubit gates: **H**, **S** and **T** gate and the two-qubit **CNOT** gate.[10]

4.8   DiVincenzo's Criteria

It may come as a surprise to some, but for a long time there existed no general, agreed upon architecture for building quantum hardware upon which quantum computations are to be executed.

In 2000 Paul DiVincenzo, working for IBM Research, published his 5 criteria [2] needed to be fulfilled to implement a quantum computer. They helped create a structural framework for quantum computing research and while it still remains open which technology will establish itself as the dominant approach to building quantum hardware, DiVincenzo's criteria are to this day the best way we can describe a quantum computer.

The criteria are:

---

[10] While the **S** gate can be created from two **T** gates, Nielsen and Chuang include the **S** gate in their universal set due to its important role in quantum error correction protocols. This is, however, beyond the scope of this book.

(C1) A scalable physical system with well characterized qubits
(C2) The ability to initialize the state of the qubits to a simple fiducial state, such as $|000...\rangle$
(C3) Long relevant decoherence times, much longer than the gate operation time
(C4) A "universal" set of quantum gates
(C5) A qubit-specific measurement capability

Criteria (C1) defines that we need to be able to characterize the qubit as two-state quantum system, with clearly distinguishable states $|0\rangle$ and $|1\rangle$. This is particularly important because it is possible to build quantum computers where the subsystems employed to act as qubits may enter other states beyond the two required by the definition of the qubit. It is then the responsibility of the computer's control apparatus to actively prevent that.

Though not explicitly mentioned, within criteria (C1), the scalability definition implies that the architecture of a quantum chip must not prohibit adding additional qubits or, alternatively, needs to allow connecting other chips to permit the steady evolution and growth of a quantum computer.

Criteria (C2) states that newly allocated qubits must be initialized to the known state. This is rather obvious, given that unitary evolution of the system according to the program's circuit must begin from a specific expected state. Additionally, quantum error correction techniques require a constant, fast supply of new qubits in well-defined state. The qubit initialization times should also be small compared to decoherence times and gate operation times.

In (C3), DiVincenzo refers to the time after which the state amplitudes of the qubits "decay". Decoherence is used to describe the dynamics of a qubit when it interacts with its surrounding. In other words, after certain period of time, typically through contact with their external environment, qubits begin to drift away from their expected state. Longer decoherence times are needed to guarantee the overall stability of the system and is absolutely critical to successful implementation of error-correction schemes.

In Sect. 4.7 we briefly discussed gate universality from the theoretical perspective. Similarly, DiVincenzo defines in (C4) that quantum computers needs to offer a set of universal gates which can be used to build universal quantum algorithms. Due to the difference in physical realization of quantum hardware technologies, this is not a straightforward proposition. In particular, various platforms would have different constraints, error rates, gate operation time and decoherence times—all of which contributing to a fragmented technological landscape. It is imperative for the success of quantum computing that algorithm authors can rely on the ability of using a standard set of gates in their work.

Finally (C5) addresses the quality and dependability of the measurement processes. DiVincenzo states that ideal fidelity is not an absolute necessity—in absence of other major impediments, the *quantum efficiency* of the measurement, so its success rate, needs to exceed 90% for quantum computing to be viable. It is also key that single shot measurements are possible.

In addition to the quantum computational criteria, DiVincenzo also laid out two additional criteria, specifically targeted toward quantum communications.

(C6)   The ability to interconvert stationary and flying qubits
(C7)   The ability to faithfully to transmit flying qubits between specified locations

In (C6) DiVincenzo reminds that while qubits within a quantum processing unit can be realized using a wide range of different physical embodiments, qubits within the theory of quantum communications are required to be transmitted between spatially separated locations, and are thus limited to very few realization possibilities—primarily photon-based. This naturally requires a reliable mechanism of conversion between these different representations.

In (C7), DiVincenzo warns that successful cross-encoding between "computational qubits" and "transmission qubits" is only the tip of the iceberg. He raises concerns related to the quality of transmissions, decoherence during the movement of qubits and points at the need of great advances required in that space, for example in the area of quantum repeaters.

References

1. Bernhardt, C. (2019). *Quantum computing for everyone*. MIT Press.
2. DiVincenzo, D. P., & IBM. (2000). The physical implementation of quantum computation. *Protein Science, 48*, 771–783.
3. Kollmitzer, C., Schauer, S., Rass, S., & Rainer, B. (2020). *Quantum random number generation*. Springer Nature Switzerland.
4. Nielsen, M. A., & Chuang, I. (2010). *Quantum computation and quantum information: 10th anniversary edition*. Cambridge University Press.
5. Rieffel, E. G., & Polak, W. H. (2014). *Quantum computing: A gentle introduction*. MIT Press.
6. Sutor, R. S. (2019). *Dancing with Qubits: How quantum computing works and how it can change the world*. Packt Publishing.

Chapter 5
Entanglement

Earlier in this book we covered some high-level quantum mechanical aspects of entanglement, with the mention of the EPR thought experiment and the refinement work done by David Bohm. We then discussed Bell's theorem and the dramatic consequences that can be drawn from it.

In Sect. 4.6, we went through various controlled gates, however we conveniently side stepped their relationship with entanglement—instead, we focused on computational basis states and merely hinted that controlled gates can also be used to create entanglement. We are now in a position to put all these pieces together and learn how entanglement arises from the mathematical formalism of quantum theory and see how it can be utilized in quantum computing.

5.1 Bell States

In classical computing based on the von Neumann architecture, COPY is one of the fundamental instructions, and the ability to copy information from one location to another is an intrinsic property of the involved hardware. Computers constantly shuttle data around various memory addresses and CPU registers to be able to perform calculations.

If we closely study the state transitions enacted by the **CNOT** gate on the computational basis states, which we outlined in (4.67), it becomes apparent that **CNOT**, in a classical sense, can also considered a simple single-bit copy machine.

In particular, if we consider only the situations when the target qubit is initialized to $|0\rangle$, which in such scheme would be considered a *blank* state, **CNOT** provides a single-bit-like copy mechanism of control state onto target. In the case when control qubit is $|0\rangle$, it copies $|0\rangle$ onto $|0\rangle$, while when control qubit is $|1\rangle$, it copies $|1\rangle$ onto $|0\rangle$.

$$|00\rangle \rightarrow |00\rangle \qquad |10\rangle \rightarrow |11\rangle \tag{5.1}$$

© The Author(s), under exclusive license to Springer Nature Switzerland AG 2022 133
F. Wojcieszyn, *Introduction to Quantum Computing with Q# and QDK*, Quantum Science and Technology, https://doi.org/10.1007/978-3-030-99379-5_5

Fig. 5.1 Bell circuit

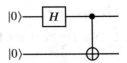

Of course this is consistent with the no-cloning theorem, which precludes copying of *unknown* quantum states, but allows copying of orthogonal states. In Sect. 2.7 we have proven that in quantum mechanics there cannot exist a universal copying transformation. We are now going to supplement this by showing how entanglement arises from an attempt to use **CNOT** on a system where the control qubit is in a superposition with respect to the standard basis.

Let us consider the quantum circuit shown on Fig. 5.1, consisting of two gates—**CNOT** and **H**.

This innocently looking two-qubit circuit, called commonly the *Bell circuit*, has profound importance in the quantum information theory. In this particular embodiment—as we will learn soon, there are three other ones as well—it is starting with both qubits in state $|0\rangle$. In this circuit, the first qubit passes through the **H** gate, which forces it into a state of a uniform linear superposition of $|0\rangle$ and $|1\rangle$. Then, a **CNOT** transformation is applied to both of the qubits, with the one in superposition acting as the control.

It is easy to calculate the output of the circuit.

$$\mathbf{CNOT}(\mathbf{H}|0\rangle \otimes |0\rangle) = \mathbf{CNOT}\left(\left(\frac{1}{\sqrt{2}}(|0\rangle + |1\rangle)\right) \otimes |0\rangle\right) = \tag{5.2}$$

$$\mathbf{CNOT}\left(\frac{1}{\sqrt{2}}|00\rangle + \frac{1}{\sqrt{2}}|10\rangle\right) = \frac{1}{\sqrt{2}}|00\rangle + \frac{1}{\sqrt{2}}|11\rangle = \frac{1}{\sqrt{2}}\left(|00\rangle + |11\rangle\right)$$

The resulting state is one of the Bell states, commonly denoted as $|\Phi^+\rangle$

$$|\Phi^+\rangle = \frac{1}{\sqrt{2}}\left(|00\rangle + |11\rangle\right) \tag{5.3}$$

We previously introduced the Bell states in Sect. 2.6 as an orthonormal basis for two qubit system. Bell states, due to entanglement, are not factorizable into the a tensor product of the individual qubit states.

Notice that in $|\Phi^+\rangle$ the probability amplitudes related to measuring the states $|01\rangle$ or $|10\rangle$ are both equal to 0, so upon measurement the system can only ever end up in a state $|00\rangle$ and $|11\rangle$, even when the individual qubits are measured independently of each other. This implies that if one of the qubits is measured to be $|0\rangle$, the second is guaranteed to be $|0\rangle$ too, while if the first one produces $|1\rangle$ on measurement, the other is certain to produce a measurement value of $|1\rangle$—provided the measurements are always done in the same basis, in this case in the computational basis.

Despite the fact that entanglement is a rather counter-intuitive concept, especially from the perspective of the macroscopic world, there is certain logic, or at least consistency behind it. We need to remember that the **H** gate creates a uniform super-position of states $|0\rangle$ and $|1\rangle$, and that the **CNOT** gate acts as the bit flip of the target qubit, but only if the control assumes a definite state $|1\rangle$. Since the control qubit is in a superposition, and that will not be resolved until a measurement is performed, the outcome of the application of **CNOT** needs to be deferred until then. In a twisted, albeit logically consistent way, since the control qubit is not in any of the definite basis states $|0\rangle$ and $|1\rangle$, the bit flip on the target qubit is both *applied and not applied* at the same time. This state of maximal uncertainty is resolved only when one of the qubits—irrespective whether it is control or target, is measured, because at this point the wave function of the system collapses and based on the probabilities calculated using the Born rule, the state of the other object in the pair also becomes known (though it still needs to be measured!).

In the example we used, Bell state $|\Phi^+\rangle$ emerges after running the so-called Bell circuit—a sequence of $\mathbf{H} \otimes \mathbf{I}$ and **CNOT** transformations—on an input pair of qubits in the initial state $|00\rangle$. Should the initial state be different—there are three additional possibilities $|01\rangle$, $|10\rangle$ and $|11\rangle$—the Bell circuit still creates entangled pairs, but corresponding to three other Bell states.

$$\mathbf{CNOT}(\mathbf{H} \otimes \mathbf{I})|00\rangle = \frac{1}{\sqrt{2}}\left(|00\rangle + |11\rangle\right) = |\Phi^+\rangle \tag{5.4}$$

$$\mathbf{CNOT}(\mathbf{H} \otimes \mathbf{I})|10\rangle = \frac{1}{\sqrt{2}}\left(|00\rangle - |11\rangle\right) = |\Phi^-\rangle$$

$$\mathbf{CNOT}(\mathbf{H} \otimes \mathbf{I})|01\rangle = \frac{1}{\sqrt{2}}\left(|01\rangle + |10\rangle\right) = |\Psi^+\rangle$$

$$\mathbf{CNOT}(\mathbf{H} \otimes \mathbf{I})|11\rangle = \frac{1}{\sqrt{2}}\left(|01\rangle - |10\rangle\right) = |\Psi^-\rangle$$

In order to create a Bell state in Q# one simply has to execute the circuit depicted in Fig. 5.1, which uses only the gates we already covered in previous chapters. For example, Listing 5.1 shows how to create the state $|\Phi^+\rangle$.

```
use (control, target) = (Qubit(), Qubit());

H(control);
CNOT(control, target);
// at this point the pair is entangled
```
Listing 5.1 Bell circuit in Q#.

Very conveniently, Q#, being a higher level quantum programming language, does not force us to only work with the bare bone concept of gates. Instead, it also offers a shortcut convenience operation to prepare a Bell state automatically. This operation, PrepareEntangledState (left : Qubit[], right : Qubit[]), is available in the Microsoft.Quantum.Preparation namespace and allows to create an arbitrary number of entangled EPR pairs. The way the API is constructed, the two arrays that are passed into the operation must contain the same amount of qubits, and the ones in the first array will become control qubits for the target qubits located at the same index position in the second array. Thus, one could replace the manual usage of **H** and **CNOT** gates with the call shown in Listing 5.2.

```
use (control, target) = (Qubit(), Qubit());

PrepareEntangledState([control], [target]);
// at this point the pair is entangled
```

Listing 5.2 Bell circuit in Q# using the built-in standard library feature instead of manual gate interaction.

We can of course test all of this in a Q# program. An operation, utilizing various features of Q# that we covered so far is shown in Listing 5.3. Since the effects of measuring a Bell state are probabilistic, we need to rerun the test code a larger number of times, and, just like we did it earlier in a book, a sample size of 4096 is chosen here.

```
operation TestBellState(init : Bool[], bases : Pauli[]) : Unit
{
    mutable res = [0, 0, 0, 0];

    for run in 1..4096 {
        use (control, target) = (Qubit(), Qubit());
        ApplyP(init[0] ? PauliX | PauliI, control);
        ApplyP(init[1] ? PauliX | PauliI, target);
        PrepareEntangledState([control], [target]);

        let c1 = resAsBool(Measure([bases[0]], [control]));
        let c2 = resAsBool(Measure([bases[1]], [target]));
        if (not c1 and not c2) { set res w/= 0 <- res[0]+1; }
        if (not c1 and not c2) { set res w/= 1 <- res[1]+1; }
        if (c1 and not c2) { set res w/= 2 <- res[2]+1; }
        if (c1 and c2) { set res w/= 3 <- res[3]+1; }
    }

    let initialState = (init[0] ? "1" | "0") + (init[1] ? "1" | "
        0");
    Message($"Initial state: |{initialState}>, measurement of
        control in {bases[0]} and of target in {bases[1]}");
    Message($"|00>: {res[0]}");
    Message($"|01>: {res[1]}");
    Message($"|10>: {res[2]}");
    Message($"|11>: {res[3]}");
}
```

Listing 5.3 Orchestration code for testing Bell states in Q#.

The testing operation allows providing input parameters that will determine the initial state of the two-qubit system—$|00\rangle$, $|01\rangle$, $|10\rangle$ or $|11\rangle$, as well as the measurement bases for both the control and target qubits. We shall initially focus on the standard basis measurements only, but the ability to reuse the same base code for other measurement bases will come in handy soon as well.

The code keeps track of the four possible measurement results using a dedicated results array, where index 0 is linked to measuring $|00\rangle$, index 1 to $|01\rangle$ and so on. After all, we are interested to check how many times the measurements of control qubit and target qubit agree with each other, which of course allows us to verify whether we really encounter the entanglement phenomenon and whether we can spot any statistical patterns in the experiment outcomes. Since Q# is immutable by default, the array that tracks results must be flagged as mutable.

For the initial state $|00\rangle$, and a measurement in the Z-basis for both qubits, the operation can now be invoked in the following manner:

```
@EntryPoint()
operation Main() : Unit {
    TestBellState([false, false], [PauliZ, PauliZ]);
}
```
Listing 5.4 Invoking the code from Listing 5.3 for initial state $|00\rangle$.

The produced output will be probabilistic, but the only measurement results should be 00 or 11, and the distribution of results should be close to the expected 50% for each.

```
Initial state: |00>, measurement of control in PauliZ and of
    target in PauliZ
Measured |00>: 2059
Measured |01>: 0
Measured |10>: 0
Measured |11>: 2037
```
Listing 5.5 Sample output from running the code from Listing 5.4.

The way the operation in 5.3 has been written, it can be easily reused to verify other combinations of input states and measurement bases. For example, the code in Listing 5.6 invokes the test operation for all four Bell states and measures the involved qubits in the computational basis.

```
@EntryPoint()
operation Main() : Unit {
    TestBellState([false, false], [PauliZ, PauliZ]);
    TestBellState([false, true], [PauliZ, PauliZ]);
    TestBellState([true, false], [PauliZ, PauliZ]);
    TestBellState([true, true], [PauliZ, PauliZ]);
}
```
Listing 5.6 Invoking the code from Listing 5.3 for all the possible two qubit initial states.

An interesting result can be observed for states $|\Psi^+\rangle$ and $|\Psi^-\rangle$, as those are correlated negatively. They can be obtained via the Bell circuit for input states $|10\rangle$ and $|11\rangle$ respectively. For example, the test operation output for $|\Psi^+\rangle$ should resemble the output shown in Listing 5.7.

```
Initial state: |01>, measurement of control in PauliZ and of
    target in PauliZ
Measured |00>: 0
Measured |01>: 2054
```

```
Measured |10>: 2042
Measured |11>: 0
```

Listing 5.7 Sample output from running the code from Listing 5.6.

Finally, it is also possible to use the test operation to verify the effects of measuring an entangled pair in mutually unbiased bases. For example, this could correspond to the hypothetical scenario where Alice measured her particle along the z-axis, and Bob, unaware of that, measured his particle along the x-axis. This is shown in Listing 5.8, explicitly using the Bell state $|\Phi^+\rangle$, though the reasoning would be identical to the other ones as well.

```
@EntryPoint()
operation Main() : Unit {
    TestBellState([false, false], [PauliZ, PauliX]);
}
```

Listing 5.8 Invoking the code from Listing 5.3 for initial state $|00\rangle$, with measurement of each qubit in a different basis.

In such situation, we should expect the results of measurement to be completely random, because the measurement in the Z-basis creates maximum uncertainty for the measurement in the X-basis. So a measurement of one qubit, let us say the control, along the z-axis, produces an eigenstate of Z-basis as measurement result. At the same time, for state $|\Phi^+\rangle$, the second qubit is guaranteed to produce the same eigenstate if it was measured in the Z-basis too. However, it is measured in the X-basis, so upon measurement it will collapse with 50% probability to $|+\rangle$ or $|-\rangle$, because the two bases are mutually unbiased. Running the code from Listing 5.6 should produce output similar to that below, experimentally confirming our reasoning.

```
Initial state: |00>, measurement of control in PauliZ and of
    target in PauliX
Measured |00>: 1006
Measured |01>: 1050
Measured |10>: 1032
Measured |11>: 1008
```

Listing 5.9 Sample output from running the code from Listing 5.6.

5.2   Testing Bell's Theorem

In Sect. 2.9 we covered Bell's theorem. In what is undoubtedly a spectacular testament to the progress in computational physics, while it took almost two decades until Bell's original idea could be verified experimentally, we can now easily test Bell's inequalities using a quantum computer, writing a program in high-level programming language like Q#.

5.2.1 Bell's Inequality in Q#

We originally introduced Bell's inequality in Eq. (2.106) as

$$Pr(\mathbf{a^+}, \mathbf{b^+}) + Pr(\mathbf{b^+}, \mathbf{c^+}) \geq Pr(\mathbf{a^+}, \mathbf{c^+}) \tag{5.5}$$

and then derived its quantum mechanical version in Eq. (2.112), which took the form of

$$\sin^2\left(\frac{\theta_{ab}}{2}\right) + \sin^2\left(\frac{\theta_{bc}}{2}\right) \geq \sin^2\left(\frac{\theta_{ac}}{2}\right) \tag{5.6}$$

We also know already that this inequality is predicted to be maximally violated when we set θ to $\frac{\pi}{3}$, forming a set of vectors as shown in Fig. 2.6.

With that in mind, we can now write a simple Q# program that will verify the violation of Bell's inequality. The overall approach that we will employ is outlined on the quantum circuit on Fig. 5.2.

The circuit takes the input qubits in the default $|00\rangle$ state and creates an entangled pair forming the singlet $|\Psi^-\rangle$ state. This is achieved by initially creating a state $|11\rangle$ and then by executing the Bell circuit. What follows is the arbitrary gate $\mathbf{U}$ which will execute the rotation of the qubits by the angles prescribed by Bell's inequality. Since we need to measure the results for $Pr(\mathbf{a^+}, \mathbf{b^+})$, $Pr(\mathbf{a^+}, \mathbf{c^+})$ and $Pr(\mathbf{b^+}, \mathbf{c^+})$, three separate unique implementations of $\mathbf{U}$ will be required. Finally, since Z-basis measurement is not appropriate in this case, a measurement in the X-basis is done at the end of the circuit. There exists no uniformly established convention for visualization of an X-basis measurement on quantum circuits—the measurement node always refers to Z-basis measurement. However, an $\mathbf{H}$ transformation followed by a standard basis measurement is a de-facto X-basis measurement, so that aspect is also visualized this way.

The internals of $\mathbf{U}$ needed to measure $Pr(\mathbf{a^+}, \mathbf{b^+})$ is shown in Fig. 5.3, while the other two vector pairs $Pr(\mathbf{a^+}, \mathbf{c^+})$ and $Pr(\mathbf{b^+}, \mathbf{b^+})$ are covered by Figs. 5.4 and 5.5 respectively. They are all simple z-axis rotations, and differ by rotation angle only—two of them create a $\frac{\pi}{3}$, while the other a $\frac{2\pi}{3}$ relative rotation angle between the two qubits.

Fig. 5.2 General circuit for testing Bell inequality

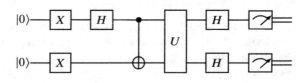

Fig. 5.3 U implementation for measuring $P(\mathbf{a^+}, \mathbf{b^+})$

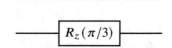

Fig. 5.4 U implementation
for measuring $P(\mathbf{a}^+, \mathbf{c}^+)$

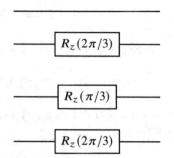

$$R_z(2\pi/3)$$

Fig. 5.5 U implementation
for measuring $P(\mathbf{b}^+, \mathbf{c}^+)$

$$R_z(\pi/3)$$

$$R_z(2\pi/3)$$

When writing Q# code, it will be most appropriate to follow similar pattern as outlined on the circuit Fig. 5.2 and build the three variants of **U** gates as standalone operations. They can then benefit from being orchestrated together in a common way, which will make the Q# code tidy and concise.

The first operation, shown in Listing 5.10, will be called Uab (q1 : Qubit, q2: Qubit) and will correspond to $Pr(\mathbf{a}^+, \mathbf{b}^+)$. It will use $\theta = \frac{\pi}{3}$ angle between **a** and **b** as outlined on Fig. 5.3. With that in mind, the corresponding Q# code is extremely simple—it only executes the rotation of the second qubit about the $|1\rangle$ state by $\frac{\pi}{3}$ using the R1 (theta : Double, qubit : Qubit) operation from Microsoft.Quantum.Intrinsic namespace. While the first qubit remains intact, the operation creates a $\frac{\pi}{3}$ offset between the two qubit states.

```
operation Uab(q1 : Qubit, q2: Qubit) : Unit {
    R1(PI() / 3.0, q2);
}
```
Listing 5.10 Q# operation for $Pr(\mathbf{a}^+, \mathbf{b}^+)$.

In a similar fashion, we name the second operation Uac (q1 : Qubit, q2: Qubit), and it will be our way to capture data for $Pr(\mathbf{a}^+, \mathbf{c}^+)$. This is corresponding to Fig. 5.4. Just like the previous one, it would transform only the second qubit, but using a larger rotation angle—because the angle between **a** and **c** should be $\frac{2\pi}{3}$. It is shown in Listing 5.11

```
operation Uac(q1 : Qubit, q2: Qubit)  : Unit {
    R1(2.0 * PI() / 3.0, q2);
}
```
Listing 5.11 Q# operation for $Pr(\mathbf{a}^+, \mathbf{c}^+)$.

Finally, operation Ubc (q1 : Qubit, q2: Qubit) from Listing 5.12 allows us to calculate $Pr(\mathbf{b}^+, \mathbf{c}^+)$. It will be the only operation where both qubits are rotated—the first one by $\frac{\pi}{3}$ while the second one by $\frac{2\pi}{3}$. The delta angle still remains $\frac{\pi}{3}$, as prescribed by Fig. 5.5 and the vector layout shown in Fig. 2.6.

```
operation Ubc(q1 : Qubit, q2: Qubit)  : Unit {
    R1(PI() / 3.0, q1);
    R1(2.0 * PI() / 3.0, q2);
}
```
Listing 5.12 Q# operation for $Pr(\mathbf{b}^+, \mathbf{c}^+)$.

With these three operations implemented we are well positioned to tackle the problem of experimental verification of Bell's inequalities. What we still need is the orchestration code that will facilitate capturing the measurements over a larger sample size. This is not much different from any of the orchestration code snippets we already wrote in earlier parts of this book.

We will call the orchestration operation Run (op: ((Qubit, Qubit) => Unit)), and its input will be a delegate to either of the three Bell's inequality **U** operation implementations. This will allow us to execute the code for each of the three variants—$Pr(\mathbf{a^+}, \mathbf{b^+})$, $Pr(\mathbf{a^+}, \mathbf{c^+})$ and $Pr(\mathbf{b^+}, \mathbf{c^+})$ and ultimately calculate the inequality.

To stay consistent with the code written so far, we shall set the sample size to 4096 iterations. Similarly to how we did in the Q# code used in Sect. 5.1, we will use a four element mutable array for keeping track of how many measurement outcomes of $|00\rangle$, $|01\rangle$, $|10\rangle$ and $|11\rangle$ we encountered. For each iteration, we allocate two fresh qubits and entangle them into the $|\Psi^-\rangle$ state by executing the Bell circuit on a pair of qubits in the initial state $|11\rangle$—the **H** gate on the first qubit, and then a **CNOT** transformation over both of the involved qubits. Next, the results for a given run are obtained via a measurement of both qubits in the X-basis. A helper function IsResultOne(input : Result) from the Microsoft.Quantum.Canon namespace is used to avoid manual comparison of the measurement result to the keyword literals Zero and One.

Because in Q# Zero results maps to the +1 eigenvalue, and because we are interested in cases where both qubits measure the +1 eigenvalue only, the final probability for $Pr(\mathbf{a^+}, \mathbf{b^+})$, $Pr(\mathbf{a^+}, \mathbf{c^+})$ and $Pr(\mathbf{b^+}, \mathbf{c^+})$ can be calculated as the amount of times we measured $|00\rangle$, divided by the total amount of iterations. IntAsDouble(a : Int) from Microsoft.Quantum.Convert namespace is needed to cast the integer representation of the running total and iterations into a double-precision floating-point number, so that the produced probability is not rounded to the integer.

```
operation Run(op: ((Qubit, Qubit) => Unit)) : Double {
    mutable res = [0, 0, 0, 0];

    let runs = 4096;
    for i in 0 .. runs {
        use (q1, q2) = (Qubit(), Qubit());
        PrepareBellState(q1, q2);
        op(q1, q2);
        let (r1, r2) = (MResetX(q1), MResetX(q2));

        if (IsResultZero(r1) and IsResultZero(r2)) {
            set res w/= 0 <- res[0]+1;
        }
        if (IsResultZero(r1) and IsResultOne(r2)) {
            set res w/= 1 <- res[1]+1;
        }
        if (IsResultOne(r1) and IsResultZero(r2)) {
            set res w/= 2 <- res[2]+1;
        }
        if (IsResultOne(r1) and IsResultOne(r2)) {
```

```
            set res w/= 3 <- res[3]+1;
        }
    }

    let p = IntAsDouble(res[0]) / IntAsDouble(runs);
    return p;
}
```

Listing 5.13 Q# orchestration code for testing Bell's inequality.

Finally, to aggregate the overall results, the orchestration operation must be invoked three times—for $Pr(\mathbf{a}^+, \mathbf{b}^+)$, $Pr(\mathbf{a}^+, \mathbf{c}^+)$ and $Pr(\mathbf{b}^+, \mathbf{c}^+)$. The final result is obtained using the Bell's formula from Eq. (5.5) as shown in the Q# code Listing 5.14.

```
@EntryPoint()
operation Main() : Unit {
    let pAB = Run(Uab);
    Message($"P(a+,b+) = "
        + DoubleAsStringWithFormat(pAB, "N2"));

    let pBC = Run(Ubc);
    Message($"P(b+,c+) = "
        + DoubleAsStringWithFormat(pBC, "N2"));

    let pAC = Run(Uac);
    Message($"P(a+,c+) = "
        + DoubleAsStringWithFormat(pAC, "N2"));

    let bellResult = pAB + pBC >= pAC;
    Message($"Bell's inequality satisfied? {bellResult}");
}
```

Listing 5.14 Invoking the code from Listing 5.13 for all three test cases.

The formatting feature used there is the `DoubleAsStringWithFormat` (a : Double, fmt : String) function from the `Microsoft.Quantum.Convert` namespace. It allows formatting floating point numeric value according to the predefined format, which is modeled after the similar feature in C#. In this particular case, using N2 means that we would want to convert the input double into a string, and display it using two decimal places.

When running the program, we should see results similar to the ones below.

```
P(a+,b+) = 0.12
P(b+,c+) = 0.12
P(a+,c+) = 0.38
Bell's inequality satisfied? False
```

Listing 5.15 Sample output from the code from Listing 5.14.

This is exactly the conclusion we came to in Eq. (2.116) while studying Bell's theorem as part of Chap. 2. The experimental outcome is clear—Bell's inequality is violated and no local hidden variable theory can reproduce quantum mechanical predictions, which also indirectly confirms the non-local nature of quantum phenomena. While we knew this all along based on the algebraic calculations done here and earlier in the book, it is nevertheless very welcome to be able confirm that we did not make any mistakes in our reasoning. In fact, taking a step back, and considering how difficult it was to experimentally confirm the Bell's theorem, it is quite mind boggling that we can simply write less than a hundred lines of Q# and quickly get the answer to the question that has troubled some of the greatest mind in the history of physics.

5.2.2 CHSH Inequality in Q#

Using similar approach as we just went through for the Bell's inequality, we can also verify the CHSH inequality. We covered the theory behind CHSH inequality in Sect. 2.9.2 and we introduced it in Eq. (2.121) as

$$\langle S_{QM} \rangle = E(\mathbf{a}, \mathbf{b}) + E(\mathbf{a}, \mathbf{b}') + E(\mathbf{a}', \mathbf{b}) - E(\mathbf{a}', \mathbf{b}') \tag{5.7}$$

$$-2 \le \langle S \rangle \le 2$$

The primary change needed is that it requires usage of four vectors, instead of three and hence we will need to define four dedicated test operations in Q# code. The CHSH vectors are arranged as shown in Fig. 2.7, namely with a $\frac{\pi}{4}$ angle between each neighboring pair.

The four required operations will be constructed using similar approach to how we constructed Q# code in Listings 5.10–5.12, namely by considering the first vector from the bottom right to be fixed, and not rotate the qubit representing it at all, while rotating the remaining three qubits by $\frac{\pi}{4}$, $\frac{\pi}{2}$ and $\frac{3\pi}{4}$ respectively. The operations are shown in Listing 5.16.

```
operation Ua1b1(q1 : Qubit, q2: Qubit) : Unit {
    R1(PI() / 4.0, q1);
    R1(2.0 * PI() / 4.0, q2);
}

operation Ua1b2(q1 : Qubit, q2: Qubit)  : Unit {
    R1(PI() / 4.0, q1);
}

operation Ua2b1(q1 : Qubit, q2: Qubit)  : Unit {
    R1(3.0 * PI() / 4.0, q1);
    R1(2.0 * PI() / 4.0, q2);
}

operation Ua2b2(q1 : Qubit, q2: Qubit)  : Unit {
    R1(3.0 * PI() / 4.0, q1);
}
```

Listing 5.16 Four operations preparing the two qubits according to CHSH inequality test defined in Fig. 2.7.

Because Q# does not allow non-alphanumeric characters in operation names, instead of naming the operations after the original vector names—**a**, **a′**, **b**, **b′**—we use the a1, a2, b1, b2 notation.

The orchestration code will be almost identical to the Listing 5.13, with the notable difference that we need to operate on expected value rather than probabilities. Therefore, instead of merely adding the probabilities for the outcomes we are interested in, we also substract the probabilities for the outcomes we are not interested in, as shown in Listing 5.17, which contains the orchestration code RunCHSH (op: ((Qubit, Qubit) => Unit)).

```
operation RunCHSH(op: ((Qubit, Qubit) => Unit)) : Double {
    mutable res = [0, 0, 0, 0];

    let runs = 4096;
    for i in 0 .. runs {
        use (q1, q2) = (Qubit(), Qubit());
        PrepareBellState(q1, q2);
        op(q1, q2);
        let (r1, r2) = (MResetX(q1), MResetX(q2));

        if (IsResultZero(r1) and IsResultZero(r2)) {
            set res w/= 0 <- res[0]+1;
        }
        if (IsResultZero(r1) and IsResultOne(r2)) {
            set res w/= 1 <- res[1]+1;
        }
        if (IsResultOne(r1) and IsResultZero(r2)) {
            set res w/= 2 <- res[2]+1;
        }
        if (IsResultOne(r1) and IsResultOne(r2)) {
            set res w/= 3 <- res[3]+1;
        }
    }

    let p = IntAsDouble(res[0] + res[3] - res[2] - res[1]) /
        IntAsDouble(runs);
    return p;
}
```

Listing 5.17 Q# orchestration code for CHSH inequality test.

Finally, we will need to invoke the orchestration operation and calculate the expectation value for the four vector pairs—(**a**, **b**), (**a**, **b′**), (**a′**, **b**) and (**a′**, **b′**). The protocol here is identical to what we did earlier in Listing 5.14.

```
@EntryPoint()
operation Main() : Unit {
    let eA1B1 = RunCHSH(Ua1b1);
    Message($"E(a1,b1) = "
        + DoubleAsStringWithFormat(eA1B1, "N2"));

    let eA1B2 = RunCHSH(Ua1b2);
    Message($"E(a1,b2) = "
```

```
      + DoubleAsStringWithFormat(eA1B2, "N2"));

  let eA2B1 = RunCHSH(Ua2b1);
  Message($"E(a2,b1) = "
      + DoubleAsStringWithFormat(eA2B1, "N2"));

  let eA2B2 = RunCHSH(Ua2b2);
  Message($"E(a2,b2) = "
      + DoubleAsStringWithFormat(eA2B2, "N2"));

  let chshResult =
      AbsD(eA1B1 + eA1B2) + AbsD(eA2B1 - eA2B2) <= 2.0;
  Message($"CHSH inequality satisfied? {chshResult}");
}
```

Listing 5.18 Invoking Q# orchestration code for the four different CHSH vector pair rotations.

The resulting CHSH inequality expects local realist hidden variable theories to stay within the range between minus two and two, as defined in Eq. (5.7). To make this check simpler we can actually compare against absolute values, which is what the code does here, utilizing the `AbsD(a : Double)` function from the `Microsoft.Quantum.Math` namespace.

Running the CHSH inequality test code prepared this way should produce output similar to the one shown in Listing 5.19, which is of course confirming our expectations.

```
E(a1,b1) = -0.69
E(a1,b2) = -0.72
E(a2,b1) = -0.73
E(a2,b2) = 0.71
CHSH inequality satisfied? False
```

Listing 5.19 Sample output of Q# CHSH inequality verification code.

5.3 Non-local Games

One gets quickly confronted with the stunning predictions of quantum mechanics when playing logic games based on Bell's inequality. In this section we will explore two examples of this remarkable category of the so-called *non-local* games.[1] In such games, typically built around simple boolean logic problems, completely against common sense, quantum mechanics, due to its non-local correlations, can produce winning strategies higher than any classical approach. The concept of *Tsirelson bound* places a mathematical restriction on the correlation between quantum mechanical objects and provides a limit by which entangled quantum objects can violate the Bell inequalities. This in turns defines by how much quantum strategies can outperform their classical counterparts in non-local games.

---

[1] An excellent overview of a wide range of such games can be found in [6].

These games are designed to be executed using quantum communications protocols, where both players are spatially separated from each other. Nevertheless, it is also possible to reconstruct them using a program executed on a quantum computer, where the game plays itself out locally, and where logical separation of the player roles happens in the code. This provides a very easy mechanism for learning and analyzing the quantum mechanical principles involved.

5.3.1 CHSH Game

The origin of the CHSH game is unclear, but most likely the first mention comes from the series of lectures by Boris Tsirelson in the 1990s [16]. The game starts with two participating players, Alice and Bob, receiving a random bit each from an impartial referee. We shall denote those bits with i_A (input Alice) and i_B (input Bob). They then proceed to randomly select a bit on their own, which we shall refer to as o_A (output Alice) and o_B (output Bob) respectively.

The goal of the game is for the players to satisfy the following formula

$$i_A \cdot i_B = o_A \oplus o_B \tag{5.8}$$

without communicating with each other during the game. They can, however, agree on the strategy to follow upfront.

The major difficulty the players need to overcome in the game is that Alice should select the most optimal value of o_A, without knowing o_B, and Bob should select best possible value of o_B, without knowing o_A, such that their selections, when the XOR logic is applied to them, would match the product of the original input bits.

Let us break down the possible development scenarios for the game. Since two bits can have four possible combinations, the randomly generated game input bits i_A and i_B could be combined to calculate the product in four different ways. Similarly, there are four different possibilities for taking an XOR between the two output bits o_A and o_B. All of those combinations are shown in Table 5.1.

By looking at the table, we can quickly realize that the most optimal classical strategy for Alice and Bob is blatantly simple. Since $i_A \cdot i_B$ produces 0 in 75% of cases, they should aim at producing 0 out of their $o_A \oplus o_B$ operation too. This means that they should both always output 0 or always 1, regardless of the input values of i_A and i_B—as long as they both stick to the same agreed value all the time. This will

Table 5.1 Possible configurations for a product and XOR taken between pairs of classical bits

i_A	i_B	$i_A \cdot i_B$	o_A	o_B	$o_A \oplus o_B$
0	0	0	0	0	0
0	1	0	0	1	1
1	0	0	1	0	1
1	1	1	1	1	0

ensure that $o_A \oplus o_B$ always results in 0, and since we already established that $i_A \cdot i_B$ gives 0 in three out of four cases, they will also win the game 75% of the time. This is the best they can do classically.

Remarkably, due to the phenomenon of entanglement and the EPR correlations, quantum strategy can lead to a higher winning probability, namely about 85%. Let us now outline the quantum approach, based on the entangled pair of qubits in the Bell state $|\Psi^-\rangle$, shared by Alice and Bob.

$$|\Psi^-\rangle = \frac{1}{\sqrt{2}}\Big(|01\rangle - |10\rangle\Big) \tag{5.9}$$

The advantage of the quantum strategy for the CHSH game is that Alice and Bob can rely on the quantum correlations in their EPR pair to help them select o_A and o_B bits in the most optimal fashion—in a way that will not be reproducible classically. This is facilitated by a scheme based on different measurement bases, carefully chosen in such a way that they can tilt the winning probability to their favor.

The strategy of Alice is not particularly complex, and can be summarized as: Alice toggles between the Z and X-bases, depending on the value of the input bit i_A, and measures her part of the Bell state qubit accordingly. If her measurement result, in either of the two bases, maps to a classical 0, she returns $o_A = 0$, otherwise it would be $o_A = 1$. The strategy is fully outlined in the Table 5.2.

As a consequence of Alice's measurements, and due to the usage of $|\Psi^-\rangle$ and its negative correlations, the quantum state of Bob's part of the entangled pair would collapse to the opposite state from what Alice measured, should he use the same measurement basis. The problem is—of course—he would not know which of the two bases Alice used, because he does not know which referee bit she received and the players are not allowed to communicate.

Here is where the unusual features of quantum correlations come into play. The players can take advantage of the fact that superposition is basis-dependent, so when a qubit in a definite state in one basis, it would be in some superposition in a different basis. The key observation to make here, is that superposition does not necessarily have to be uniform. Instead, it may be more likely to produce one of the eigenstates over the other, for example:

$$|\psi\rangle = \frac{4}{5}|0\rangle + \frac{3}{5}|1\rangle \tag{5.10}$$

Table 5.2 Alice strategy for the CHSH game

Alice input bit	$i_A = 0$		$i_A = 1$	
Alice basis	Z		X	
Alice measurement result	$|0\rangle$	$|1\rangle$	$|+\rangle$	$|-\rangle$
Alice output bit	$o_A = 0$	$o_A = 1$	$o_A = 0$	$o_A = 1$

In this particular case the probability amplitudes are uneven, so according to the Born rule, the probability of measuring $|0\rangle$ would be $|\frac{4}{5}|^2$ which is 64%, while $|1\rangle$ would be $|\frac{3}{5}|^2$ which is 36%.

In addition to that, we already know that Z and X-bases are mutually unbiased, so measuring in the X-basis a qubit that is known to be in a eigenstate $|0\rangle$ or $|1\rangle$ of the Z-basis, produces measurement results with maximum uncertainty. However instead of measuring by arbitrarily selecting one these commonly used bases, Bob can try to choose his measurement basis close enough to Z and X, in such a way that his probability amplitude for a collapse to the eigenstate (and ultimately eigenvalue) opposite of Alice, is as large as possible when Alice measured her qubit, and use that trick to optimally select his o_B.

It turns out that the most optimal value for Bob in the CHSH game is to use the basis that is rotated from the Z-basis by $\theta = \frac{\pi}{8}$ clockwise and anti-clockwise around the y-axis. We will call the first basis $\frac{\pi}{8}$-basis, with eigenvectors represented in the standard basis are

$$|u_1\rangle = \cos\left(\frac{\pi}{8}\right)|0\rangle + \sin\left(\frac{\pi}{8}\right)|1\rangle \qquad (5.11)$$

$$|u_2\rangle = -\sin\left(\frac{\pi}{8}\right)|0\rangle + \cos\left(\frac{\pi}{8}\right)|1\rangle$$

We will call the other basis $-\frac{\pi}{8}$-basis, with eigenvectors

$$|v_1\rangle = \cos\left(-\frac{\pi}{8}\right)|0\rangle + \sin\left(-\frac{\pi}{8}\right)|1\rangle \qquad (5.12)$$

$$|v_2\rangle = -\sin\left(-\frac{\pi}{8}\right)|0\rangle + \cos\left(-\frac{\pi}{8}\right)|1\rangle$$

The strategy for Bob is then summarized in Table 5.3.

Of course this does not answer the real question yet—why would such specific basis choice and quantum strategy work better than its classical counterpart?

Imagine that Alice receives $i_A = 0$ from the referee, measures in the Z-basis, receives a measurement outcome $|0\rangle$ and therefore outputs $o_A = 0$. Such measurement result would mean Bob's qubit is now collapsed to $|1\rangle$, because they are negatively correlated in the Bell state.

Since Alice produced $o_A = 0$, for them to win the game, Bob needs to output $o_B = 0$ too, something that he will do—according to Table 5.3—if he manages to measure $|u_2\rangle$ or $|v_2\rangle$. Notice that these states, being "second eigenstates" should really map to a classical 1, but an additional consequence of the negative correlation

Table 5.3 Bob strategy for the CHSH game

Bob input bit	$i_B = 0$		$i_B = 1$	
Bob basis	$\frac{\pi}{8}$		$-\frac{\pi}{8}$	
Bob measurement result	$|u_1\rangle$	$|u_2\rangle$	$|v_1\rangle$	$|v_2\rangle$
Bob output bit	$o_B = 1$	$o_B = 0$	$o_B = 1$	$o_B = 0$

of the initial entangled state $|\Psi^-\rangle$ is that Bob will need to return his output bit values o_B that are opposite to his classical measurement results.

Suppose now that Bob receives $i_B = 0$ from the referee, so he would be measuring in the $\frac{\pi}{8}$ basis. By inspecting the states from Eq. 5.11, we can see that for the input referee bits $i_A = 0$, $i_B = 0$, the winning combination of states are (Alice's state first, then Bob's) $|0, u_2\rangle$ and $|1, u_1\rangle$, which the players will encounter with the following probabilities

$$Pr_{win1} = |\langle 0, u_2 | \Psi^- \rangle|^2 = \frac{1}{2} \cos^2\left(\frac{\pi}{8}\right) \tag{5.13}$$

$$Pr_{win2} = |\langle 1, u_1 | \Psi^- \rangle|^2 = \frac{1}{2} \cos^2\left(\frac{\pi}{8}\right)$$

By adding them we find out that Bob will be able to observe the desired measurement result, and thus produce the winning bit o_B to pair with o_A, with a total probability of $cos^2(\frac{\pi}{8}) \approx 85\%$.

Following similar reasoning we could easily calculate this to be the same for all the other permutations of the game input conditions.

We can implement the CHSH game in Q# in just a few simple steps, which will naturally help verify our theoretical discussion. If the reasoning we went through above is correct, we should indeed see a classical strategy providing 75% winning rate, with the quantum strategy outperforming it with its 85% success rate.

Let us first look at the game code shown in Listing 5.20. The operation RunGame (strategy : (Bool, Bool) => (Bool, Bool)) will take care of drawing the random referee bits, as well as invoke the provided strategy—be it classical or quantum. The strategy can be passed in as a delegate, which accepts two referee input bits arguments, and which returns two game output bits.

```
operation RunGame(strategy : (Bool, Bool) => (Bool, Bool)) :
    Double {
    let runs = 4096;
    mutable wins = 0;

    for i in 0..runs {
        let inputA = DrawRandomBool(0.5);
        let inputB = DrawRandomBool(0.5);

        let chosenBits = strategy(inputA, inputB);
        if ((inputA and inputB) == Xor(chosenBits)) {
            set wins += 1;
        }
    }

    return IntAsDouble(wins) / IntAsDouble(runs);
}
```

Listing 5.20 CHSH game strategy testing code in Q#.

Notice that we also expanded the terse mathematical naming scheme used so far, like i_A or i_B, to more descriptive variable names such as inputA or inputB. For the

purposes of code readability and verbosity, this pattern will be repeated throughout
the rest of the Q# implementation.

The code executes a loop of 4096 shots, which is the general sample size
we have been using throughout this book so far. The referee bits inputA and
inputB for Alice and Bob are drawn using the operation that is part of Q# standard library—DrawRandomBool (successProbability : Double) from the
Microsoft.Quantum.Random namespace. It is a very convenient helper operation that generates a quantum-backed random bit value. The parameter allows us to
control the statistical distribution of the retrieved values—passing in a 0.5 means
that we will get a truly random value.

The results of the execution of the game strategy, which are returned as tuples
of booleans, are compared to the randomly drawn referee bits using the CHSH
game formula from Eq. (5.8). Q# actually provides a handy built-in function Xor (a
: Bool, b : Bool) in the Microsoft.Quantum.Logical namespace which we
use to verify the outcome of the game. We keep the running total of wins in a mutable
integer variable, and at the end of the orchestration operation, we calculate the win
percentages for the strategy.

Let us now turn our attention to the implementations of both classical and quantum
strategies. The classical one is very straightforward—as we already know, returning
$o_A = 1$ and $o_B = 1$ or $o_A = 0$ and $o_B = 0$ will result in a 75% success rate. Listing 5.21 shows such simple implementation, where both o_A and o_A are returned as 0
bit values. Notice that the lack of quantum features means that it could be declared as
a Q# function rather than a Q# operation, however, because our game delegate
from Listing 5.20 requires a strategy to be supplied as operation, we define an
operation here as well.

```
operation ClassicalStrategy(inputA : Bool, inputB : Bool)
    : (Bool, Bool) {
    return (false, false);
}
```

Listing 5.21 CHSH game classical strategy implementation in Q#.

Next, we need to have a look at the implementation of quantum strategy, which
is shown in Listing 5.22.

```
operation QuantumStrategy(inputA : Bool, inputB : Bool)
    : (Bool, Bool) {
    use (qubitA, qubitB) = (Qubit(), Qubit());
    X(qubitA);
    X(qubitB);
    H(qubitA);
    CNOT(qubitA, qubitB);

    let resultA = MeasurementA(inputA, qubitA);
    let resultB = not MeasurementB(inputB, qubitB);
    return (resultA, resultB);
}
```

Listing 5.22 CHSH game quantum strategy implementation in Q#.

The operation QuantumStrategy (inputA : Bool, inputB : Bool) allocates two qubits for both of the players and initializes the entangled Bell state $|\Psi^-\rangle$. It then returns two bits—represented by two booleans—corresponding to the result of Alice's and Bob's quantum measurement outcomes. Because of the negative correlation between qubits in the entangled state $|\Psi^-\rangle$, the result of Bob's measurement must be negated, hence the usage of the not operator in the code.

As we already outlined, Alice and Bob will measure their qubits in different ways, depending on the input bits received from the referee, so we need to define two dedicated Q# operations to capture the logic relevant for each of them separately. They are called MeasurementA (bit : Bool, q : Qubit) and MeasurementB (bit : Bool, q : Qubit) in the code. The implementation of Alice's quantum strategy—the simpler one of the two—is shown in Listing 5.23.

```
operation AliceMeasurement(bit : Bool, q : Qubit) : Bool {
    let result = Measure([bit ? PauliX | PauliZ], [q]);
    return IsResultOne(result);
}
```
Listing 5.23 Alice's measurement procedure in the CHSH game.

Alice will measure in the Z-basis when the bit she received from the referee is $i_A = 0$, and she will use the X-basis when $i_A = 1$. In this particular case the operation Measure(bases : Pauli[], qubits : Qubit[]) allows us to dynamically select the measurement basis for the qubit by passing in the relevant keyword literal based on the referee bit.

Finally, let's explore Bob's strategy, implemented in Q# in Listing 5.24.

```
operation BobMeasurement(bit : Bool, q : Qubit) : Bool {
    let rotationAngle = bit ? (2.0 * PI() / 8.0) | (-2.0 * PI() /
        8.0);
    Ry(rotationAngle, q);
    return IsResultOne(M(q));
}
```
Listing 5.24 Bob's measurement procedure in the CHSH game.

Bob would need to measure in the $\pm\frac{\pi}{8}$ basis, but there is no built-in Q# feature for measurements in an arbitrary basis like that one. However, just like the X-basis is the Z-basis rotated around the y-axis by $-\frac{\pi}{4}$, the $\pm\frac{\pi}{8}$ bases measurement can be reconstructed by the usage of an appropriate rotation about the y-axis followed by a regular Z-basis measurement. One thing we need to remember is that while the orthogonal vectors for real numbers are $\frac{\pi}{2}$ apart, on the Bloch sphere, the orthogonal vectors point to the antipodal points on the surface of the sphere, so they are π radians apart. Due to that, we need to multiply the desired rotation angle by a factor of 2. This is the reason why in the code we rotate not by (+/-)PI() / 8.0, but instead of by (+/-)2.0 * PI() / 8.0.

This is almost everything that was needed to complete the program, however we still need to add the Q# program entry point which will invoke the game code from Listing 5.20 for both classical and quantum strategies, as well as output the results. This operation is shown in Listing 5.25.

```
@EntryPoint()
operation Main() : Unit {
    let classicalWinRate = RunGame(ClassicalStrategy);
    Message($"Classical win probability: "
        + DoubleAsStringWithFormat(classicalWinRate * 100., "N2")
            );

    let quantumWinRate = RunGame(QuantumStrategy);
    Message($"Quantum win probability: "
        + DoubleAsStringWithFormat(quantumWinRate * 100., "N2"));
}
```

Listing 5.25 Q# code for CHSH game orchestration.

The output from running the program should confirm our expectations and resemble the following:

```
Classical win probability: 75.95
Quantum win probability: 84.74
```

Listing 5.26 Sample output from the CHSH game code from Listing 5.25.

One final wrinkle that can be added to the code is to check if it would make any difference to the outcome of the experiment, if we tried to randomize the measurement order in the quantum strategy. So far, due to how the code was written—sequentially—Alice's measurement are always performed first. This can be tested by making a small modification to the code from Listing 5.22.

```
operation QuantumStrategy(inputA : Bool, inputB : Bool)
    : (Bool, Bool) {
    use (qubitA, qubitB) = (Qubit(), Qubit());
    X(qubitA);
    X(qubitB);
    H(qubitA);
    CNOT(qubitA, qubitB);

    let shouldAliceMeasureFirst = DrawRandomBool(0.5);
    if (shouldAliceMeasureFirst) {
        let resultA = MeasurementA(inputA, qubitA);
        let resultB = not MeasurementB(inputB, qubitB);
        return (resultA, resultB);
    }
    else
    {
        let resultB = not MeasurementB(inputB, qubitB);
        let resultA = MeasurementA(inputA, qubitA);
        return (resultA, resultB);
    }
}
```

Listing 5.27 CHSH game quantum strategy implementation in Q#, with randomized measurement order between Alice and Bob.

In the updated variant of quantum strategy, we introduced an extra random boolean variable that is drawn with success probability of 50%, to determine whether Alice should perform her measurement first. The remaining times, Bob will run his measurements, based on the $\pm\frac{\pi}{8}$ bases first, and Alice would follow him. As it turns out, however, due the bidirectional nature of quantum correlations, such randomization of the process would not have any impact on the outcome.

This is exactly what we expected based on the theory of quantum computing. In a rather remarkable outcome, by replacing classical bits with entangled qubits, without any classical communication, the CHSH game receives a ten percentage points boost over the best logical classical winning scenario. We all know that in sports passion often trumps logic, but as the Bell's theorem and EPR correlations show, quantum mechanics also has a way of trumping (classical) logic.

5.3.2 GHZ Game

The violations of Bell's inequality show up as statistical results of the Bell's test when performed a larger number of times and involving spatially separated observers. As such, while the conclusions we draw from it, once the experiment is analyzed, are rather astonishing, the overall process of determining the incompatibility with local realism is certainly somewhat convoluted and cumbersome as it relies on data from statistical ensembles.

As it turns out, the so-called *GHZ state* allows us to experience a violation of local hidden variables theories using a single shot experimental setup. In other words, instead of illustrating the conflict with local realism with probabilistic results, the GHZ state can do that with deterministic approach as well. The GHZ state was first proposed in 1989 by Daniel Greenberger, Michael Horne and Anton Zeilinger [9], who suggested to go beyond using the usual EPR pairs, as it is the case in Bell's theorem, and employ systems consisting of larger amounts of particles when trying to prove that no local hidden variables theory could be used to reproduce the predictions of quantum theory. This drastically simplifies the reasoning, compared to using the usual Bell states.

The GHZ state is a maximally entangled three qubit state that has the following form:

$$|GHZ\rangle = \frac{1}{\sqrt{2}}\left(|000\rangle + |111\rangle\right) \tag{5.14}$$

It is a special three qubit state of a more general class of multi-qubit entangled states that take the following form:

$$|\psi\rangle = \frac{1}{\sqrt{2}}\left(|0\rangle^{\otimes n} + |1\rangle^{\otimes n}\right) \tag{5.15}$$

We can create a GHZ state using the same steps as needed to create the state $|\Phi^+\rangle$, followed by an extra **CNOT** between the second qubit making up the $|\Phi^+\rangle$ pair and an extra third qubit. This can be expressed algebraically as:

$$|GHZ\rangle = (\mathbf{I} \otimes \mathbf{CNOT})(\mathbf{CNOT} \otimes \mathbf{I})(\mathbf{H} \otimes \mathbf{I} \otimes \mathbf{I})|000\rangle \qquad (5.16)$$

The corresponding quantum circuit is depicted on Fig. 5.6.

Let us now imagine the following logical game. A referee generates three random bits x, y and z and hands one bit to each of the three players. In return, each player is supposed to provide a single bit back to the referee—we will call them a, b and c. The goal of the players to satisfy the following logical condition:

$$a \oplus b \oplus c = x \vee y \vee z \qquad (5.17)$$

Similarly to the CHSH game, the players cannot communicate with each other once they receive their bits, however they can agree on a common strategy upfront. While the referee could issue $2^3 = 8$ possible combinations of x, y and z, the game only restricts referee to choose from one of the four predefined sets of x, y and z outlined in Table 5.4. The table also lists the corresponding winning condition which must be fulfilled by the players.

A quick analysis of the task would tell us that it is logically impossible for the players to win the game 100% of the time, and that the best classical strategy can be 75% successful. That would be a strategy where all players, regardless of their input bit, return 1 as their output.

Remarkably, when using a GHZ state, the players can win the game every single time. To achieve that, the players will need to map the two possible states of the referee

Fig. 5.6 Circuit to create a GHZ state

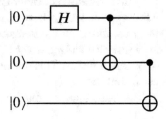

Table 5.4 Combinations of referee bits in GHZ game, and the winning condition required by the players to produce

x	y	z	Winning condition
0	0	0	0
0	1	1	1
1	0	1	1
1	1	0	1

bit to unitary transformations shown in (5.18), and apply the relevant transformation
to their specific part of the GHZ triplet prior to measuring it in the standard basis.

$$0 \rightarrow \mathbf{H} \tag{5.18}$$
$$1 \rightarrow \mathbf{HS}^\dagger$$

Let us first consider the case when the referee issues $x = 0$, $y = 0$ and $z = 0$.
Based on the mapping in (5.18), the players—each one independently of course, and
not knowing what bit their co-players received—will apply the $\mathbf{H}$ transformation to
the qubit in their possession. Altogether, this changes the original $|GHZ\rangle$ state to
the following:

$$(\mathbf{H} \otimes \mathbf{H} \otimes \mathbf{H})|GHZ\rangle = \frac{1}{2}\left(|000\rangle + |011\rangle + |101\rangle + |110\rangle\right) \tag{5.19}$$

This state clearly indicates that when each of the qubits making up the $|GHZ\rangle$
triplet is transformed by the $\mathbf{H}$ transformation and then measured in the computational
basis, the players altogether have (based on the Born rule) an equal 25% probability
of having the entire system collapse to $|000\rangle$, $|011\rangle$, $|101\rangle$ or $|110\rangle$.

Irrespective of which of these four states ends up being the measurement outcome,
when we now take the XOR of each of the variations, we find out that they always
produce a 0, satisfying the first of the winning conditions.

$$0 \oplus 0 \oplus 0 = 0 \quad 0 \oplus 1 \oplus 1 = 0 \quad 1 \oplus 0 \oplus 1 = 0 \quad 1 \oplus 1 \oplus 0 = 0 \tag{5.20}$$

Let us now consider another situation, this time when the referee distributes $x = 1$,
$y = 1$ and $z = 0$. Based on (5.18), the players will now need to perform a transformation:

$$(\mathbf{HS}^\dagger \otimes \mathbf{HS}^\dagger \otimes \mathbf{H})|GHZ\rangle = \frac{1}{2}\left(|001\rangle + |010\rangle + |100\rangle + |111\rangle\right) \tag{5.21}$$

Such composite set of transformations—consisting of one independently executed
by each player—would now create a different set of possible measurement outcomes,
again with 25% each, namely, $|001\rangle$, $|010\rangle$, $|100\rangle$ or $|111\rangle$. The XOR for each possible
outcome is always equal to 1.

$$0 \oplus 0 \oplus 1 = 1 \quad 0 \oplus 1 \oplus 0 = 1 \quad 1 \oplus 0 \oplus 0 = 1 \quad 1 \oplus 1 \oplus 1 = 1 \tag{5.22}$$

This, yet again, satisfies the rules of the game and puts the players in the winning
position. The other two cases of referee bits would follow similar reasoning and we
shall skip them—however the players would also win the game as long as they stick
to the transformations prescribed by the mapping (5.18).

All in all, the GHZ state enables the players to always win the game, something
that is impossible using classical logic only.

The key to understanding how GHZ state is able to provide correlations stronger than classical is to realize, as diligent reader may have already done, that the two transformations used in the GHZ game, $\mathbf{H}$ and $\mathbf{HS}^\dagger$, when followed by a default Z-basis measurement, are nothing more than measurements done in the X and Y-bases respectively. This leads us to the heart of the mystery surrounding the GHZ state.

Let us first consider the case where all three players measure in the X-basis. We can construct the corresponding observable-operator $\mathbf{X} \otimes \mathbf{X} \otimes \mathbf{X}$ where $\mathbf{X}$ is the Pauli matrix σ_x, and see what are its effects on the $|GHZ\rangle$. Note that we are not performing a measurement here, but instead using this operator as a regular unitary transformation to see where it would lead us. When applied to the GHZ state $|GHZ\rangle$ from Eq. (5.14) it gives us

$$(\mathbf{X} \otimes \mathbf{X} \otimes \mathbf{X})|GHZ\rangle = \begin{bmatrix} 0\,0\,0\,0\,0\,0\,0\,1 \\ 0\,0\,0\,0\,0\,0\,1\,0 \\ 0\,0\,0\,0\,0\,1\,0\,0 \\ 0\,0\,0\,0\,1\,0\,0\,0 \\ 0\,0\,0\,1\,0\,0\,0\,0 \\ 0\,0\,1\,0\,0\,0\,0\,0 \\ 0\,1\,0\,0\,0\,0\,0\,0 \\ 1\,0\,0\,0\,0\,0\,0\,0 \end{bmatrix} \frac{1}{\sqrt{2}} \begin{bmatrix} 1 \\ 0 \\ 0 \\ 0 \\ 0 \\ 0 \\ 0 \\ 1 \end{bmatrix} = |GHZ\rangle \qquad (5.23)$$

This leads us to a surprising conclusion. It tells us that $|GHZ\rangle$ is the eigenstate of an operator $\mathbf{X} \otimes \mathbf{X} \otimes \mathbf{X}$ with the eigenvalue of $+1$. In other words, if all three players measure their parts of the tripartite entangled GHZ state in the X-basis, the product of their measurement results is going to be $+1$. If we now denote measurement outcomes of players A, B and C as A_x, B_x and C_x (since measurements happen in the X-basis), we can conclude that

$$A_x \cdot B_x \cdot C_x = 1 \qquad (5.24)$$

To avoid repetition, we will spare ourselves from performing this calculation for the other three observable-operators that can be used by the players in the game—those where two of the components of the GHZ triplet are measured in the Y-basis (using the σ_y observable)—but that can easily be done in a similar way.[2] It would lead us to the following outcome

$$(\mathbf{X} \otimes \mathbf{Y} \otimes \mathbf{Y})|GHZ\rangle = -|GHZ\rangle \qquad (5.25)$$
$$(\mathbf{Y} \otimes \mathbf{X} \otimes \mathbf{Y})|GHZ\rangle = -|GHZ\rangle$$
$$(\mathbf{Y} \otimes \mathbf{Y} \otimes \mathbf{X})|GHZ\rangle = -|GHZ\rangle$$

which in turn means that $|GHZ\rangle$ is the eigenstate of each of these three observable-operators with the eigenvalue of -1. This tells us if two players measure their parts

---

[2] In general, going forward in the book we will avoid expanding the kets to classic matrix format to avoid the unnecessary verbosity.

of the tripartite entangled GHZ state in the Y-basis, while one of them measures in the X-basis, the product of their measurement results is going to be -1.

Just as we did it before, we can summarize this using measurement outcomes A, B and C:

$$A_x \cdot B_y \cdot C_y = A_y \cdot B_x \cdot C_y = A_y \cdot B_y \cdot C_x = -1 \qquad (5.26)$$

This also means that

$$(A_x \cdot B_y \cdot C_y) \cdot (A_y \cdot B_x \cdot C_y) \cdot (A_y \cdot B_y \cdot C_x) = -1 \qquad (5.27)$$

However one can easily point out that no matter whether $+1$ or -1 was measured the following is true

$$A_y^2 = 1 \qquad B_y^2 = 1 \qquad C_y^2 = 1 \qquad (5.28)$$

We can use that to rewrite Eq. (5.27) as

$$A_x \cdot B_x \cdot C_x = -1 \qquad (5.29)$$

This is of course the exact opposite from the expected value of $+1$ we calculated in Eq. (5.24). If local hidden variable theory replicating the predictions of quantum mechanics was possible, it would have meant that both of these incompatible values of $A_x \cdot B_x \cdot C_x$ would have needed to be "pre-encoded" into the system simultaneously. This is of course blatantly contradicting the very idea behind such theory. Instead, in quantum mechanics the product of measurement outcomes $A_x \cdot B_x \cdot C_x$ assumes the value of either $+1$ or -1 contextually, depending on the basis choices done by participating (independent!) observers.

The GHZ puzzle illustrates, in a far more dramatic fashion, what was already established in earlier sections by analyzing the Bell's and CHSH inequalities. Quantum mechanics seems to completely upend the picture of what our human common sense considers to be a reasonable model of reality, one that the EPR trio tried to assert in their original paper.

Let us start the Q# implementation of the GHZ game by defining the classical function that will simulate the best possible classical strategy. Since we already established that there is no complicated logic behind it, the code is very simple too— we will simply return 1, represented by a boolean *true*, for each of the three player bits a, b and c. In that sense the function is very similar to the classical strategy used in the CHSH game in Listing 5.21.

```
function ClassicalStrategy(x : Bool, y : Bool, z : Bool)
  : Bool[] {
    return [true, true, true];
}
```

Listing 5.28 Classical strategy for the GHZ game implemented in Q#.

Since the quantum strategy will use the GHZ state, we will need to include the transformations necessary to create it, according to the quantum circuit defined in Fig. 5.6. Then each of the three qubits will be measured either in the PauliY basis (when the referee bit is 1) or PauliX basis (when the referee bit is 0). At the end of the operation, similarly to how it was done in the classical strategy, all three bits will be returned from the operation in a form of an array.

```
operation QuantumGHZStrategy(x : Bool, y : Bool, z : Bool) : Bool
    [] {
    use qubits = Qubit[3];
    H(qubits[0]);
    CNOT(qubits[0], qubits[1]);
    CNOT(qubits[1], qubits[2]);

    let basis1 = x ? PauliY | PauliX;
    let basis2 = y ? PauliY | PauliX;
    let basis3 = z ? PauliY | PauliX;

    let result = [Measure([basis1], [qubits[0]]) == One,
                  Measure([basis2], [qubits[1]]) == One,
                  Measure([basis3], [qubits[2]]) == One];

    return result;
}
```

Listing 5.29 Quantum strategy for the GHZ game implemented in Q#.

The next piece of the puzzle is going to be the check for the game outcome. This will follow the rules specified in Eq. (5.17) and is shown in Listing 5.30.

```
function CheckGameOutcome(xyz : Bool[], abc : Bool[]) : Bool {
    return Xor(abc[0], Xor(abc[1], abc[2])) == (xyz[0] or xyz[1]
        or xyz[2]);
}
```

Listing 5.30 GHZ game result check implemented in Q#.

Finally, we need the orchestration code which will bring all the pieces together. The four sets of referee input bits can be hardcoded as an array of four boolean arrays. For each of these inputs, we shall execute both quantum and classical strategies and keep track of the amount of wins achieved by each of them. In the end we will expect 100% win success for the quantum approach, and 75% win probability for the classical one—more precisely the case when referee deals out $x = y = z = false$ should always result in a failure of the classical methodology. To accumulate a larger sample size, we will play the game 128 times for each of the four sets of referee bits. The entire orchestrator code is shown next.

```
@EntryPoint()
operation Main() : Unit {
    let xyz = [
        [false, false, false],
        [false, true, true],
        [true, false, true],
        [true, true, false]
```

```
];

mutable qStrategyWins = 0;
mutable cStrategyWins = 0;

for refBits in xyz {
    mutable qStrategyRoundWins = 0;
    mutable cStrategyRoundWins = 0;

    for _ in 1 .. 128 {
        let qOutput = QuantumGHZStrategy(
            refBits[0], refBits[1], refBits[2]);
        let qGameResult = CheckGameOutcome(refBits, qOutput);
        set qStrategyWins += qGameResult ? 1 | 0;
        set qStrategyRoundWins += qGameResult ? 1 | 0;

        let cOutput = ClassicalStrategy(
            refBits[0], refBits[1], refBits[2]);
        let cGameResult = CheckGameOutcome(refBits, cOutput);
        set cStrategyWins += cGameResult ? 1 | 0;
        set cStrategyRoundWins += cGameResult ? 1 | 0;
    }

    Message($"For input: {refBits} Quantum GHZ strategy won:
        {cStrategyRoundWins} times.");
    Message($"For input: {refBits} Classical strategy won: {
        cStrategyRoundWins} times.");
}

Message($"Quantum GHZ strategy won: {qStrategyWins *
    100/512}%.");
Message($"Classical strategy won  : {cStrategyWins *
    100/512}%.");
}
```

Listing 5.31 Orchestration code to run the GHZ game in Q#.

The orchestration code prints out the results as well, so upon running this program, we shall see the same console output as shown in 5.32.

```
Quantum GHZ strategy won: 100%.
Classical strategy won: 75%.
```

Listing 5.32 Expected output from running the code from Listing 5.31.

This is of course in line with our expectations, and in agreement with the predictions of quantum mechanics.

5.4 Teleportation

Aside from time travel and various forms of extraterrestrial life, teleportation is quite possibly the most spectacular concept in science fiction writing. Even more remarkably, teleportation is an integral part of quantum information theory, where it is a natural consequence of both entanglement and the linearity of quantum mechanics.

Quantum teleportation was first introduced in 1993 by Bennett and Brassard [2], who developed an ingenious idea of relying on a combination of EPR correlations and classical bits to move the complete set of information about an undetermined quantum state $|\psi\rangle$ from one physical location to another. The reason why we refer to it as *teleportation* is that the protocol does not require a physical quantum communication channel over which the data describing the teleported state travels—in the process of quantum teleportation, there is no physical spatial movement of matter or particles.

In the teleportation scheme, the description of the original *source* quantum state effectively gets destroyed in one place, and recreated in another location over a *target* quantum object. This is also the subtle difference between science-fiction and the quantum mechanical phenomenon—it is not the physical quantum object itself that gets destroyed in the original location, but the information about its quantum state—the object itself remains in existence, albeit in a different quantum state. Conversely, in the target location the teleported object does not appear out of nowhere, but instead another pre-existing quantum object assumes the informational structure of the original object.

This information destruction at the source happens via the processes of measurement, forcing the quantum object out of its $|\psi\rangle$ state and into one of the eigenstates of the measurement basis, such as for example $|0\rangle$ or $|1\rangle$ when using the standard computational basis. The state $|\psi\rangle$ is then reconstructed onto another quantum object through a clever series of entanglements. In other words, at any given time, only one of the actors at both the source and target sides of the teleportation process can reconstruct the original quantum state [14].

This is also precisely the reason why the process of teleportation does not violate the no-cloning theorem—at no point does the same quantum state exist in two places, as original and a copy.[3] Additionally, as should soon become apparent, the teleportation protocol requires a classical communication channel between the two actors to be available as well, ensuring that no faster-than-light information transmission can occur—even though the quantum information can be considered to be propagate instantaneously by means of the state collapse of the EPR pair [7].

The protocol was first successfully implemented experimentally in 1997, by two independent groups—one, using a four-photon design, lead by Anton Zeilinger at the University of Innsbruck [4], and the other, using a two-photon procedure, by Sandu Popescu at the University of Rome "La Sapienza" [3].

In 2006, a group from the Niels Bohr Institute in Copenhagen [15] reported a successful teleportation of a quantum state encoded in a laser beam, into a quantum system made up of an atomic ensemble of caesium atoms. This in itself is a phenomenal achievement, as it represented the first time a quantum state was teleported between light and matter. It also marked an important milestone in practical appli-

---

[3] While the analogy is not perfect, it may be helpful to think of teleportation as cut-and-paste, instead of copy-and-paste.

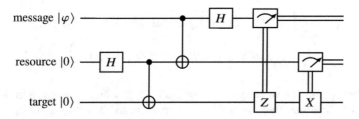

Fig. 5.7 Quantum teleportation circuit

cation of quantum information theory, which allows encoding the same information in quantum systems of different nature.

In 2014 researchers [5] demonstrated quantum teleportation of the polarization state of a telecom-wavelength photon onto the state of a solid-state quantum memory, which could underpin transferring quantum information between remote quantum networks. In 2015, a group from the University of Science and Technology of China [13], carried out a successful teleportation experiment between a ground observatory and low orbit Earth satellite—some 1400 km away.

The circuit depicting the teleportation protocol is shown in Fig. 5.7.

We can start reasoning about the teleportation protocol by imagining the usual two actors, Alice and Bob, who are spatially separated from each other and would want to teleport a qubit state $|\varphi\rangle$, that is currently in the possession of Alice. It is in an arbitrary superposition state and is labeled as *message* on the circuit.

Naturally, Alice cannot just measure her qubit and send the measurement result to Bob using classical means, because a measurement would cause the quantum state collapse and would destroy the unique quantum state—simply yielding one of the eigenstates of the measurement basis.

To facilitate the teleportation process, Alice and Bob are also sharing an entangled pair of qubits, represented by any of the four Bell states. We shall use $|\Phi^+\rangle$ as an example here, but the other three states are equally appropriate. This entangled pair is labeled on the circuit as *resource*, which denotes Alice's qubit and *target*, which represents Bob's qubit. This entangled state is, as we already know, created by applying the sequence of **H** transformation followed by a two qubit **CNOT**.

$$|\Phi^+\rangle = \frac{1}{\sqrt{2}}\Big(|00\rangle + |11\rangle\Big) \tag{5.30}$$

Taking all of this into account, Alice and Bob are dealing with a three qubit system, consisting of the message qubit $|\varphi\rangle$ that is owned by Alice, and the shared entangled pair of qubits $|\Phi^+\rangle$, where one part is in possession of Alice and the other is controlled by Bob. This overall initial system state can be annotated as $|\psi_1\rangle$, and it will be a tensor product between the message $|\varphi\rangle$ and the EPR pair $|\Phi^+\rangle$.

$$|\psi_1\rangle = |\varphi\rangle \otimes |\Phi^+\rangle = (a|0\rangle + b|1\rangle) \otimes \frac{1}{\sqrt{2}}\Big(|00\rangle + |11\rangle\Big) = \tag{5.31}$$

$$\frac{1}{\sqrt{2}}\Big(a|0\rangle \otimes (|00\rangle + |11\rangle) + b|1\rangle \otimes (|00\rangle + |11\rangle)\Big) =$$
$$\frac{1}{\sqrt{2}}\Big(a|000\rangle + a|011\rangle + b|100\rangle + b|111\rangle\Big)$$

At this point, let us recall the four Bell states which we first introduced in Eq. (2.90). Together, they form an orthonormal basis for a two-qubit system in a four dimensional Hilbert space—for two maximally entangled particles. We can therefore perform a joint measurement over two qubits in the Bell basis and obtain one of the Bell states as a result—in case of a uniform superposition, with equal probabilities.

Since all quantum transformations are reversible, we can also apply a reverse Bell circuit—**CNOT**, followed by **H**, to transform the Bell state back to the original input single qubits states. Using this principle, a Bell basis measurement can be converted into a standard measurement in the computational basis.

As a result, the next step for Alice is to allow her message qubit that will be teleported, $|\varphi\rangle$, to interact with her part of the entangled pair $|\Phi^+\rangle$. She first applies a **CNOT** transformation to the two qubits in her possession, with the qubit that she wishes to teleport acting as a control qubit.

This creates an overall system state $|\psi_2\rangle$, which we can describe this mathematically in the following way:

$$|\psi_2\rangle = (\mathbf{CNOT} \otimes \mathbf{I})|\psi_1\rangle = \frac{1}{\sqrt{2}}\Big(a|000\rangle + a|011\rangle + b|110\rangle + b|101\rangle\Big) \quad (5.32)$$

State $|\psi_2\rangle$ depicts a system "halfway" into the reverse Bell circuit, since the **CNOT** is applied already. Alice then follows with a Hadamard transformation on the message qubit $|\varphi\rangle$. Executing **H** on the first qubit in the system, using the state $|\psi_2\rangle$ as input, results in state $|\psi_3\rangle$.

$$|\psi_3\rangle = (\mathbf{H} \otimes \mathbf{I} \otimes \mathbf{I})|\psi_2\rangle = \quad (5.33)$$

$$\frac{1}{2}\Big(a(|000\rangle + |011\rangle + |100\rangle + |111\rangle) + b(|010\rangle + |001\rangle - |110\rangle - |101\rangle)\Big) =$$
$$\frac{1}{2}\Big(|00\rangle(a|0\rangle + b|1\rangle) + |01\rangle(a|1\rangle + b|0\rangle)+$$
$$|10\rangle(a|0\rangle - b|1\rangle) + |11\rangle(a|1\rangle - b|0\rangle)\Big)$$

The final picture becomes clearer now. Upon Alice performing the measurement over the pair of qubits in her possession, the wave function collapses into one of the four unentangled states—with equal probability of 25%—she could receive $|00\rangle$, $|01\rangle$, $|10\rangle$ or $|11\rangle$. As we see in state $|\psi_3\rangle$ in Eq. (5.33), each of these results has a direct impact on the state of the qubit of Bob. Those are summarized in Table 5.5.

Table 5.5 Overview of the teleportation protocol

Alice measurement	Bob qubit state	Bob normalized state
$|00\rangle$	$a|0\rangle + b|1\rangle$	$a|0\rangle + b|1\rangle$
$|01\rangle$	$a|1\rangle + b|0\rangle$	$\mathbf{X}(a|0\rangle + b|1\rangle)$
$|10\rangle$	$a|0\rangle - b|1\rangle$	$\mathbf{Z}(a|0\rangle + b|1\rangle)$
$|11\rangle$	$a|1\rangle - b|0\rangle$	$\mathbf{ZX}(a|0\rangle + b|1\rangle)$

As a consequence of Alice's measurement, Bob's qubit can assume one of four possible states. It either already is in the correct state $|\varphi\rangle$, or is in the state $|\varphi\rangle$ transformed by $\mathbf{X}, \mathbf{Z}$ or $\mathbf{ZX}$, which are of course the single qubit Pauli transformations.

Consider the first case listed in Table 5.5, where Alice measurement produced two eigenstates $|00\rangle$. In that case Bob's qubit is in the state $|\varphi\rangle$ and we can consider the teleportation as completed, since that is really the original state of Alice's qubit. In the three other situations, Bob would still to apply additional corrective transformations on the target qubit in order to successfully recover the original state $|\varphi\rangle$.

Because Pauli gates are unitary and Hermitian, Bob can rely on them to reproduce the original quantum message state $|\varphi\rangle$. All he needs to do is to simply apply either $\mathbf{X}, \mathbf{Z}$ or $\mathbf{ZX}$, in order to normalize his qubit's state at the desired $|\varphi\rangle$

$$\mathbf{X}(a|1\rangle + b|0\rangle) = a|0\rangle + b|1\rangle = |\varphi\rangle \tag{5.34}$$

$$\mathbf{Z}(a|0\rangle + b|1\rangle) = a|0\rangle + b|1\rangle = |\varphi\rangle$$

$$\mathbf{ZX}(a|1\rangle - b|0\rangle) = a|0\rangle + b|1\rangle = |\varphi\rangle$$

While all of that, in quite a stunning fashion, allows Bob to extract the initial quantum message state that he and Alice wanted to be teleported, the problem is that, of course, Bob does not know the state without measuring it. As a result, Bob does not know which of the three transformations to apply to it, or whether to apply any at all.[4] This problem is solved in the teleportation protocol by requiring an additional mandatory classical communication channel, such as a phone connection, to exist between Alice and Bob. This could be any channel, as long as classical bits can reliably flow through it. Alice would then communicate to Bob the result of her measurement to Bob using 2 classical bits—00, 01, 10 or 11. Bob knows that his four corrective transformations are tied to Alice's measurement result and thus can choose the relevant Pauli transformation to recover the qubit state based on that information.

This completes the teleportation protocol. The end result is that Alice's original quantum state $|\varphi\rangle$ on her message qubit gets destroyed upon measurement and that qubit assumes one of the orthogonal basis states $|0\rangle$ or $|1\rangle$ of the Z-basis. Alice's qubit forming the entangled pair with Bob also gets *used up*—it got measured and was forced to collapse to $|0\rangle$ or $|1\rangle$ as well. On the other hand, Bob's qubit assumed the state of $|\varphi\rangle$, $\mathbf{X}|\varphi\rangle$, $\mathbf{Z}|\varphi\rangle$ or $\mathbf{ZX}|\varphi\rangle$. Bob, based on the information about Alice's measurement results, received from her using classical means, managed to correct

---

[4] Not applying any transformation may also simply mean applying $\mathbf{I}$.

it to back to $|\varphi\rangle$. Bob can now measure that qubit, should he choose to do so, or use it in any other quantum algorithm. They can naturally repeat this procedure for any number of qubits, provided they are equipped with enough entangled pairs to facilitate the teleportation process.

The requirement of sharing the two classical bits between Alice and Bob to finalize the protocol also ensures that the teleportation procedure obeys the theory of special relativity, that is, that it cannot be used for superluminal communication. This was also pointed out by the original authors of the teleportation procedure, who assert that the process requires classical channel of two bits of capacity [2].

The methods applied in the teleportation scheme are generally referred to in quantum information theory as LOCC, *local operations and classical communication*. The concept captures the following activities of the involved parties:

- they apply local quantum mechanical unitary transformations on the quantum objects in their possession
- they use classical communication channels to announce measurement results, and those measurement results may then be used to execute further local transformations

Both aspects of LOCC are an integral part of the process, necessary for a given protocol to be successfully completed. Without any of the parts, the outcome would be random.

The process of quantum entanglement is remarkable for two primary reasons. First of all, an arbitrary quantum state got teleported using no physical quantum connection between the two actors Alice and Bob. Yes, there was a classical communication link between to transmit the two classical bits representing Alice's measurement, but no specific information regarding the quantum state was shared. Additionally, it means that the message was moved between Alice and Bob in a completely safe way—there is no way for an eavesdropper or a malicious party to intercept the quantum message. Even the concept of the *eavesdropper* is rendered obsolete—since there is nothing to *eaves drop on*—as there is no physical connection related to the movement of the quantum state. It simply does not travel through spacetime in a classical understanding of the word.

Secondly, and this is may not be immediately clear, using only two bits of classical data, Alice facilitated sending a continuous quantum state—after all the coefficients a and b could be any values as long as $|a|^2 + |b|^2 = 1$. Even a numerical description of such state would be impossible to express classically in just two bits.

Let us now reconstruct the teleportation protocol using the quantum computer—with Q# and QDK as our tooling vector. While we are free to teleport any arbitrary quantum state, to illustrate the teleportation process well, we shall define a dedicated operation that prepares a unique state as shown in Listing 5.33.

```
operation PrepareState(q : Qubit) : Unit is Adj + Ctl {
    Rx(1. * PI() / 2., q);
    Ry(2. * PI() / 3., q);
    Rz(3. * PI() / 4., q);
}
```

Listing 5.33 Preparing an arbitrary quantum state for teleportation

The operation applies some arbitrary rotations around all three axes, and defines an `Adjoint` specialization. This way, we will be able to prepare the state to be teleported starting from a computational basis eigenstate such as $|0\rangle$, teleport it, and afterwards apply the adjoint to return back to $|0\rangle$ to verify the correctness of the teleportation protocol.

When writing the Q# code for the protocol, we will use variable names corresponding to the ones we already established on the quantum circuit—message, resource and target, to facilitate better cognitive association between the circuit and the code.

Next, we define a Q# operation to execute the large part of the logic from circuit Fig. 5.7, and which shall be our vessel to execute the protocol. The operation is shown in Listing 5.34, where we call it `Teleport` (message : `Qubit`, resource : `Qubit`, target : `Qubit`), and the three qubits it accepts as parameters will be the ones over which the teleportation procedure will play out. The message will already come in prepared in the state it should be teleported in, while the other two qubits, one representing Alice, the other Bob, are passed into the operation in the default, unentangled, state.

```
operation Teleport(message : Qubit, resource : Qubit, target :
    Qubit) : Unit {
    // create entanglement between resource and target
    H(resource);
    CNOT(resource, target);

    // reverse Bell circuit on message and resource
    CNOT(message, resource);
    H(message);

    // mesaure message and resource
    let messageResult = M(message) == One;
    let resourceResult = M(resource) == One;

    // and decode state
    DecodeTeleportedState(messageResult, resourceResult, target);
}
```
Listing 5.34 Step one of the teleportation protocol in Q#.

The operation creates a Bell state using the resource and target qubits. Those are now entangled and will facilitate the teleportation process. Next, a reverse Bell circuit, to decompose a joint measurement in the Bell basis into standard single qubit measurements in the computational basis is applied on message and resource. Finally, we perform a Z-basis measurement on both message and resource mimicking the actions of Alice that we discussed earlier.

At this point Alice has two classical bits in her possession and will pass them over to Bob, who will complete the protocol and transform the received state of his qubit to decode the original teleported message state. This is achieved by calling the DecodeTeleportedState(messageResult : Bool, resourceResult : Bool, target : Qubit) operation, shown in Listing 5.35.

```
operation DecodeTeleportedState(
    messageResult : Bool,
    resourceResult : Bool,
    target : Qubit) : Unit {

    if not messageResult and not resourceResult {
        I(target);
    }
    if not messageResult and resourceResult {
        X(target);
    }
    if messageResult and not resourceResult {
        Z(target);
    }
    if messageResult and resourceResult {
        Z(target);
        X(target);
    }
}
```

Listing 5.35 Decoding step of the teleportation protocol in Q#.

We follow precisely the rules we already defined Table 5.5, so depending on the outcome of Alice's measurements, the relevant decoding transformation is applied by Bob, who invokes an **I** (no-op), **X**, **Z** or **ZX** against the target qubit. We should remember that all four possibilities in this case are equally likely to be the result of the joint Bell measurement, each one with the probability of 25%, so the decoding step required to be applied may be different from run to run.

The final thing left to do, is to define the entry point operation for our Q# program and make use of the Teleport (message : Qubit, resource : Qubit, target : Qubit) operation from Listing 5.34. As usually, it would make sense to gather results from a larger sample size, so, to ensure the outcome is always as expected, we will execute the teleportation procedure in a loop of 4096 iterations. The orchestration code also needs to take care of allocating the three required qubits upon every iteration, invoke the state preparation operation from Listing 5.33 against the message qubit, and then its adjoint against the target qubit to verify that the target really acquired the correct quantum state. The entire operation is shown in Listing 5.36.

```
@EntryPoint()
operation Start() : Unit {
    let runs = 4096;
    mutable successCount = 0;

    for i in 1..runs {
        use (message, resource, target) = (Qubit(), Qubit(),
            Qubit());
```

```
        // prepare the source state
        PrepareState(message);

        // execute teleportation
        Teleport(message, resource, target);

        // verify the target state
        Adjoint PrepareState(target);
        set successCount += M(target) == Zero ? 1 | 0;
    }

    Message("Success rate: "
        + DoubleAsStringWithFormat(
            100. * IntAsDouble(successCount) / IntAsDouble(runs),
                "N2"));
}
```

Listing 5.36 Orchestration operation for executing the teleportation code a larger number of times.

Invoking this program should yield the output like the one shown in 5.37—the reported success rate, of course when running on a simulator or when the hardware offers perfect fidelity, would be 100%.

```
Success rate: 100.00
```

Listing 5.37 Expected output from running the code from Listing 5.36.

5.5 Entanglement Swapping

Another fascinating consequence of the teleportation protocol is that it can be used to entangle two qubits that are spatially separated and never interacted with each other in the first place. This is possible by using two pairs of entangled qubits such as any of the four Bell states.

Due to the fact that entanglement swapping is effectively a variation of quantum teleportation, the circuit representing it, shown on Fig. 5.8, is very similar to the teleportation circuit that we studied in Fig. 5.7.

Imagine Alice and Bob, independently of each other, holding a pair of qubits in the state $|\Phi^+\rangle$. We have not needed that so far, but in order to not lose track which qubits are owned by Alice and which by Bob, let us additionally use a subscript-based annotation of qubits. They will be denoting which qubits belong to the pair owned by Alice—those with indices 1 and 2—and which belong to a pair owned by Bob—those with indices 3 and 4. Because of that we shall refer to both of the Bell states not as $|\Phi^+\rangle$ but as $|\psi_a\rangle$ (for Alice) and $|\psi_b\rangle$ (for Bob).

$$|\psi_a\rangle = \frac{1}{\sqrt{2}}\left(|00\rangle_{12} + |11\rangle_{12}\right) \tag{5.35}$$

$$|\psi_b\rangle = \frac{1}{\sqrt{2}}\left(|00\rangle_{34} + |11\rangle_{34}\right)$$

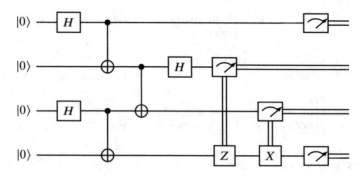

Fig. 5.8 Entanglement swapping circuit

The initial overall system state is the usual tensor product of both of these $|\psi\rangle$ pairs.

$$|\psi_a\rangle \otimes |\psi_b\rangle = \frac{1}{\sqrt{2}}\left(|00\rangle_{12} + |11\rangle_{12}\right) \otimes \frac{1}{\sqrt{2}}\left(|00\rangle_{34} + |11\rangle_{34}\right) = \qquad (5.36)$$
$$\frac{1}{2}\left(|00\rangle_{12}|00\rangle_{34} + |00\rangle_{12}|11\rangle_{34} + |11\rangle_{12}|00\rangle_{34} + |11\rangle_{12}|11\rangle_{34}\right)$$

We can rearrange the order of indices slightly

$$|\psi_a\rangle \otimes |\psi_b\rangle = \qquad (5.37)$$
$$\frac{1}{2}\left(|00\rangle_{14}|00\rangle_{23} + |01\rangle_{14}|01\rangle_{23} + |10\rangle_{14}|10\rangle_{23} + |11\rangle_{14}|11\rangle_{23}\right)$$

Due to linearity of quantum mechanics, this can be further rewritten into tensor products of different Bell states.

$$|\psi_a\rangle \otimes |\psi_b\rangle = \qquad (5.38)$$
$$\frac{1}{2}\left(|\Phi^+\rangle_{14} \otimes |\Phi^+\rangle_{23} + |\Psi^+\rangle_{14} \otimes |\Psi^+\rangle_{23} + |\Psi^-\rangle_{14} \otimes |\Psi^-\rangle_{23} + |\Phi^-\rangle_{14} \otimes |\Phi^-\rangle_{23}\right)$$

Now consider Alice and Bob handing off one of their qubits to a third party, which we shall call Carol. She receives qubits with index 2 from Alice, and with index 3 from Bob, and performs a joint Bell measurement over them—similar to that shown in Eqs. (5.32) and (5.33). Depending on Carol's measurement result, the system spanning the other two qubits, with indices 1 and 4 now collapses to one of the four Bell states $|\Phi^+\rangle_{14}$, $|\Psi^+\rangle_{14}$, $|\Psi^-\rangle_{14}$ or $|\Phi^-\rangle_{14}$. Similarly to how it was done in the core teleportation flow, Alice and Bob do not know the character of their correlations yet, so Bob would need to apply corrective transformations based on Carol's results—exactly as it was done in Eq. (5.34).

All things considered, this is quite a spectacular result, because Carol, by measuring qubits indexed 2 and 3 effectively *projected* the entanglement onto qubits indexed 1 and 4, which are now entangled—even though they never interacted with each other in the first place or never shared any common history.

In a traditional set up, entanglement between two particles is obtained by allowing those particles to physically interact, for example by simultaneously emitting two photons from the same source. Entanglement swapping along the lines of the protocol steps discussed here, using autonomous continuous wave sources, was first realized experimentally in 2007 by a group at University of Geneva [10].

In order to implement entanglement swapping with Q# we will reuse most of the code from the Q# teleportation implementation and Listing 5.36. The operation for entanglement swapping will follow similar steps, except we will need to start with two entangled pairs, one belonging to Alice and one to Bob. Just as we discussed moments ago, the state $|\Phi^+\rangle$ is the easiest to use from the code perspective too.

The operation EnganglementSwapping() is shown in Listing 5.38. It will return a tuple of two classical bits as Bool values representing the results of the measurement of Alice's and Bob's qubits—those that were not handed off to Carol and that became entangled even though they never interacted in the first place.

```
operation EnganglementSwapping() : (Bool, Bool) {
    // Alice
    use (q1, q2) = (Qubit(), Qubit());
    H(q1);
    CNOT(q1, q2);

    // Bob
    use (q3, q4) = (Qubit(), Qubit());
    H(q3);
    CNOT(q3, q4);

    // rest of the protocol to follow
}
```

Listing 5.38 Initial step of the entanglement swapping protocol in Q#.

At this point our actors both hold an independently created entangled pair of qubits, denoted by (q1, q2) for Alice and (q3, q4) for Bob. In the next step, they hand off one qubit from their Bell states to Carol—Alice sends her qubit q2, while Bob shares q4. Carol proceeds to apply the reverse Bell circuit to facilitate the execution of the Bell basis measurement, similar to how we did it in regular teleportation. Code in Listing 5.39 extends the operation from Listing 5.38 with this logic.

```
// Bell measurement - done by Carol
CNOT(q2, q4);
H(q2);

let q2Result = M(q2);
let q4Result = M(q4);
```

Listing 5.39 Step two of the entanglement swapping protocol in Q#.

From this point on, the qubits measured by Carol, (q2, q4), are no longer needed and can be discarded. The result of Carol's measurement are two classical bits, which Carol hands over to Bob. The current state of the system is that the *other* qubits of the entangled pairs owned by Alice and Bob, namely qubits q1 and q3, have become entangled, but could be in any of the four Bell states.

One of Alice or Bob can now use the measurement results received from Carol to transform the qubit in his or her possession in order to change the entangled state spanning q1 and q3 into the desired $|\Phi^+\rangle$ state. Let us imagine Bob doing it—he can follow the same rules as shown in teleportation Listing 5.35.

```
if (q2Result == Zero and q4Result == Zero) {
    I(q3);
}

if (q2Result == Zero and q4Result == One) {
    X(q3);
}

if (q2Result == One and q4Result == Zero) {
    Z(q3);
}

if (q2Result == One and q4Result == One) {
    X(q3);
    Z(q3);
}
```

Listing 5.40 Step three of the entanglement swapping protocol in Q#.

The final thing we will want to do here is to measure the qubits q1 and q3 and return their classical bit results as the output of the operation. This will allow us to execute this protocol in a loop and gather some statistically significant results. The entire operation with all the steps, including this last one, collapsed into a single code block is shown in Listing 5.41.

```
operation EnganglementSwapping() : (Bool, Bool) {
    // Alice
    use (q1, q2) = (Qubit(), Qubit());
    H(q1);
    CNOT(q1, q2);

    // Bob
    use (q3, q4) = (Qubit(), Qubit());
    H(q3);
    CNOT(q3, q4);

    // Bell measurement - done by Carol
    CNOT(q2, q4);
    H(q2);

    let q2Result = M(q2);
    let q4Result = M(q4);
```

```
    if (q2Result == Zero and q4Result == Zero) {
        I(q3);
    }
    if (q2Result == Zero and q4Result == One) {
        X(q3);
    }
    if (q2Result == One and q4Result == Zero) {
        Z(q3);
    }
    if (q2Result == One and q4Result == One) {
        X(q3);
        Z(q3);
    }

    let c1 = ResultAsBool(M(q1));
    let c3 = ResultAsBool(M(q3));
    return (c1, c3);
}
```

Listing 5.41 Complete entanglement swapping protocol in Q#.

We are now only missing the simple orchestrator that will run the operation Listing 5.41 a larger amount of times and collect the output. It will execute the swapping 4096 times, and record the measurement results over q1 and q3, categorized into buckets for classical 00, 01, 10 and 11. It will also check if the measurement results for the two qubits match. The entire orchestrator is shown in Listing 5.42.

```
@EntryPoint()
operation Start() : Unit {
    mutable matchingMeasurement = 0;
    mutable res = [0, 0, 0, 0];

    for i in 1..4096 {
        let (c1, c3) = EnganglementSwapping();
        if (not c1 and not c3) { set res w/= 0 <- res[0]+1; }
        if (not c1 and c3) { set res w/= 1 <- res[1]+1; }
        if (c1 and not c3) { set res w/= 2 <- res[2]+1; }
        if (c1 and c3) { set res w/= 3 <- res[3]+1; }
        if (c1 == c3) { set matchingMeasurement += 1; }
    }

    Message($"Measured |00>: {res[0]}");
    Message($"Measured |01>: {res[1]}");
    Message($"Measured |10>: {res[2]}");
    Message($"Measured |11>: {res[3]}");
    Message($"Measurements of qubits 1 and 3 matched: {
        matchingMeasurement}");
}
```

Listing 5.42 Orchestrating the operation from Listing 5.41.

Because of the fact that qubits q1 and q3 should, at the end of the protocol, and prior to the final measurements, be in the state $|\Phi^+\rangle$, we should see perfect correlation in their measurement results, and the only measurement outcomes should be either

00 or 11, with equal probability each. This should be confirmed by the program's output, which after running, should resemble the one from 5.43.

```
Measured |00>: 2046
Measured |01>: 0
Measured |10>: 0
Measured |11>: 2050
Measurements of qubits 1 and 3 matched: 4096
```

Listing 5.43 Sample output from the code from Listing 5.42.

We need to note that this does not fully confirm that q1 and q3 are really in state $|\Phi^+\rangle$, as the states $|\Phi^+\rangle$ and $|\Phi^-\rangle$ are actually indistinguishable using measurement only, because they only differ by the phase. To better confirm that we are indeed dealing with state $|\Phi^+\rangle$ we can extend the entanglement swapping operation with an extra line of code, to be executed after Bob's fix-up of his qubit state, and prior to performing measurements of q1 and q3:

```
Adjoint PrepareEntangledState([q1], [q3]);
```

Listing 5.44 Invoking an adjoint of an operation preparing the Bell state $|\Phi^+\rangle$.

This of course relies on one of the basic postulates of quantum mechanics, namely that every operation is by definition reversible, and that applying the adjoint of a transformation will effectively undo this transformation. Given that we expect the qubits to be in state $|\Phi^+\rangle$, applying the adjoint of `PrepareEntangledState` `(left : Qubit[], right : Qubit[]) : Unit`, which in Q# creates the state $|\Phi^+\rangle$ from two qubits in state $|00\rangle$, should return those qubits back to state $|00\rangle$.

And this is indeed what we should see in this case. After such a change to the `EnganglementSwapping()` operation, we should now have both q1 and q3 in state $|0\rangle$ upon measurement, and therefore all we should see in the measurement results would be classical 00. This can be confirmed by running the code again.

```
Measured |00>: 4096
Measured |01>: 0
Measured |10>: 0
Measured |11>: 0
Measurements of qubits 1 and 3 matched: 4096
```

Listing 5.45 Sample output from the code testing entanglement swapping updated with the adjoint call from Listing 5.44.

5.6 Superdense Coding

Another canonical example of using entanglement to achieve an outcome that is beyond reach of classical physics is the so-called *superdense coding* protocol. It was first proposed by Bennett and Wiesner in 1992 [1]. The unusual name stems from the fact that the protocol allows sending two classical bits of information by physically moving only a single qubit. From that perspective, as we shall learn in this section, it can also be considered to be a conceptual inverse of teleportation.

Just like in the case of quantum teleportation, superdense coding starts of with two actors, Alice and Bob, sharing an entangled EPR pair of qubits. The source of entanglement is irrelevant—they could entangle their qubits when they are both in physical presence of one another, or the entangled pair may come from a third party source. Once the entanglement is established, Alice and Bob can be spatially separated for the rest of the protocol.

Alice decides that she would like to send Bob a classical two-bit message. Such a message can have four possible contents

$$00, 01, 10, 11 \qquad\qquad (5.39)$$

Naturally, given that the message consists of two bits, classically, Alice would need to send those two bits for Bob in order to transfer this information to him. In quantum communications, thanks to superdense coding protocol, she can convey these two bits of classical information using a single qubit that is currently in her possession.

Bennett and Wiesner realized that due to how entangled particles behave, Alice can apply one of four unique unitary transformations on her part of the entangled pair, and the result of that operation is the system evolving into one of the four orthogonal Bell states making up an orthonormal basis for two-qubit systems. Of course a two-qubit quantum system, just like a two-bit classical one, is suitable to carry two bits of information. In other words, Alice can encode two classical bits of information into the entangled pair by applying a relevant unitary transformation to her qubit *only*, even if Alice and Bob are spatially separated from each other.

Alice can then send that single qubit to Bob using a quantum channel, such as a quantum optical link. Upon a joint measurement in the Bell basis spanning the qubit he already had in his possession and the qubit received from Alice, Bob can decode the two classical bits—and deconstruct the original two-bit message encoded by Alice.

Ultimately the actors still need to leverage two qubits to send two classical bits of information, which is not much different from having two classical bits sent over between the two parties in a classical way. However in the quantum case, the initial EPR pair is shared upfront and the communication protocol is ultimately completed by physically moving only one of the qubits. The net result is two bits of information are transported when sending only one physical qubit. The other qubit, just like in the case of teleportation, assumes the correct state through the quantum processes which we cannot explain any better than through non-local effects.

From a purely theoretical standpoint, the superdense coding protocol was an important milestone in the development of quantum information theory. At the same time, its practical benefits are rather questionable, especially given the instability of quantum communication links, rapid improvements in classical data compression protocols and the quality and capacity of classical information channels, such as fiber networks.

As such, generally speaking, using bits wherever possible is still dramatically cheaper and preferred over using qubits. On top of that, in the field of quantum communications, there are still a lot of complexities related to stably moving entangled

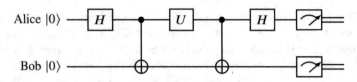

Fig. 5.9 Quantum superdense coding circuit

particles over long distances, and then storing them carefully, while preserving the cohesion of the entanglement [12].

Despite these difficulties, there are additional application scenarios for superdense coding protocol beyond the original "data compression" concept, such as in the field of quantum security. The prerequisite of physically requiring both qubits to decode the encoded message makes superdense coding a very attractive tool for secure communications—it means that even if the qubit in transit is intercepted by an eavesdropper, it alone cannot be used to recover the initial message. One such approach was suggested by a group of researches back in 2005 [17]. Superdense coding was experimentally confirmed for the first time, using a pair of entangled photons, by scientists from the University of Innsbruck in 1996 [11].[5]

As we mentioned before, superdense coding is often referred to as the inverse of the quantum teleportation protocol. The reason for this is that the decoding step that is needed in the teleportation process by the receiving party (Bob), happens to be an identical set of unitary transformation that must be applied in the encoding step the sending party (Alice) needs to apply in the superdense coding protocol.

The workflow of the protocol is illustrated by the quantum circuit Fig. 5.9. The **U** is the encoding gate that can become a Pauli **I**, **X**, **Z** or **ZX** transformation—depending on the data that Alice would like to encode.

Let us walk through this procedure step by step. Alice and Bob start off with a pre-prepared pair of maximally entangled qubits, in our case it will be the Bell state $|\Phi^+\rangle$, though other Bell states are equally suitable.

$$|\Phi^+\rangle = \frac{1}{\sqrt{2}}(|00\rangle + |11\rangle) \tag{5.40}$$

Alice and Bob can then set off in their separate directions, each one with a qubit of their own in their possession. Alice would then encode two bits of classical information into this shared entangled pair—and she can do it by applying one of the aforementioned single-qubit Pauli transformation to her qubit only. She chooses the appropriate transformation according to the rules outlined in Table 5.6.

By doing so, she may end up leaving the joint system state intact as $|\Phi^+\rangle$—in the case of the **I** transformation, or she may end up transforming the state of the entangled pair to one of the three other Bell states $|\Psi^+\rangle$, $|\Phi^-\rangle$ or $|\Psi^-\rangle$. This is summarized

---

[5] A diligent reader may recall that one of the first realizations of the teleportation protocol also happened at University of Innsbruck, and also under the leadership of Anton Zeilinger [4].

Table 5.6 Overview of the superdense coding protocol when the initial system state is $|\Phi^+\rangle$

Alice encoding	Alice transformation	System state	Bob decoding
00	**I**	$|\Phi^+\rangle$	$|00\rangle$
01	**X**	$|\Psi^+\rangle$	$|01\rangle$
10	**Z**	$|\Phi^-\rangle$	$|10\rangle$
11	**ZX**	$|\Psi^-\rangle$	$|11\rangle$

algebraically in Eq. (5.41). Note that the second qubit, physically held by Bob, is not being acted upon in any way at this stage.

$$(\mathbf{I} \otimes \mathbf{I}) \frac{1}{\sqrt{2}} \left(|00\rangle + |11\rangle \right) = \frac{1}{\sqrt{2}} \left(|00\rangle + |11\rangle \right) = |\Phi^+\rangle \qquad (5.41)$$

$$(\mathbf{X} \otimes \mathbf{I}) \frac{1}{\sqrt{2}} \left(|00\rangle + |11\rangle \right) = \frac{1}{\sqrt{2}} \left(|01\rangle + |10\rangle \right) = |\Psi^+\rangle$$

$$(\mathbf{Z} \otimes \mathbf{I}) \frac{1}{\sqrt{2}} \left(|00\rangle + |11\rangle \right) = \frac{1}{\sqrt{2}} \left(|00\rangle - |11\rangle \right) = |\Phi^-\rangle$$

$$(\mathbf{ZX} \otimes \mathbf{I}) \frac{1}{\sqrt{2}} \left(|00\rangle + |11\rangle \right) = \frac{1}{\sqrt{2}} \left(|01\rangle - |10\rangle \right) = |\Psi^-\rangle$$

At that point Alice sends off her qubit to Bob using a quantum communications channel, who is now in complete possession of the entangled pair. Bob can decode the two bits of information using a reverse Bell circuit—a **CNOT** on both qubits, followed by an **H** transformation on the qubit he received. As discussed in earlier when covering teleportation, this allows Bob to decompose a joint measurement in the Bell basis into standard single qubit measurements in the computational basis.

To be more explicit, depending on which of the four Bell states Bob is dealing with (which, in turn, depends on the encoding operation applied by Alice in the first place), as soon as Bob executes the **CNOT** transformation, the system state evolves into one the following four possibilities:

$$\mathbf{CNOT}|\Phi^+\rangle = \frac{1}{\sqrt{2}} \left(|00\rangle + |10\rangle \right) = \frac{1}{\sqrt{2}} \left(|0\rangle + |1\rangle \right) \otimes |0\rangle \qquad (5.42)$$

$$\mathbf{CNOT}|\Psi^+\rangle = \frac{1}{\sqrt{2}} \left(|11\rangle + |01\rangle \right) = \frac{1}{\sqrt{2}} \left(|1\rangle + |0\rangle \right) \otimes |1\rangle$$

$$\mathbf{CNOT}|\Phi^-\rangle = \frac{1}{\sqrt{2}} \left(|00\rangle - |10\rangle \right) = \frac{1}{\sqrt{2}} \left(|0\rangle - |1\rangle \right) \otimes |0\rangle$$

$$\mathbf{CNOT}(|\Psi^-\rangle = \frac{1}{\sqrt{2}} \left(|01\rangle - |11\rangle \right) = \frac{1}{\sqrt{2}} \left(|0\rangle - |1\rangle \right) \otimes |1\rangle$$

What follows is the **H** transformation applied to the first qubit, taking the system into its final pre-measurement state. This state would be one of four cases, that manifest themselves as:

$$(\mathbf{H} \otimes \mathbf{I}) \frac{1}{\sqrt{2}} \left(|0\rangle + |1\rangle \right) \otimes |0\rangle = |00\rangle \qquad (5.43)$$

$$(\mathbf{H} \otimes \mathbf{I}) \frac{1}{\sqrt{2}} \left(|1\rangle + |0\rangle \right) \otimes |1\rangle = |01\rangle$$

$$(\mathbf{H} \otimes \mathbf{I}) \frac{1}{\sqrt{2}} \left(|0\rangle - |1\rangle \right) \otimes |0\rangle = |10\rangle$$

$$(\mathbf{H} \otimes \mathbf{I}) \frac{1}{\sqrt{2}} \left(|0\rangle + |1\rangle \right) \otimes |1\rangle = |11\rangle$$

At the end Bob is in possession of two qubits, which, upon measurement in the Z-basis, with 100% probability collapse to the two eigenstates exactly corresponding to the two classical bits that Alice encoded, either 00, 01, 10 or 11. This concludes the protocol.

We are now ready to shift our attention towards experimental implementation of the superdense coding protocol in Q#. Our Q# code will be structured in a way that, similarly to other demos in this book, will allow for repeated execution and collection of results over a larger sample size.

We will begin by implementing both the encoding and the decoding operation according to the circuit from Fig. 5.9. Encoding will mimic the steps Alice went through in our theoretical discussion and accepts two classical data bits and a single qubit as arguments—Encode (value1 : Bool, value2 : Bool, qubit : Qubit). This operation is shown in Listing 5.46.

```
operation Encode(value1 : Bool, value2 : Bool, qubit : Qubit) :
    Unit {
    if not value1 and not value2 {
        I(qubit);
    }
    if not value1 and value2 {
        X(qubit);
    }
    if value1 and not value2 {
        Z(qubit);
    }
    if value1 and value2 {
        X(qubit);
        Z(qubit);
    }
}
```

Listing 5.46 Encoding operation of the superdense coding protocol.

Depending on which two-bit message Alice would like to send to Bob, the encoding procedure chooses a no-op **I** transformation or one of the Pauli gate combinations—**X**, **Z** or **ZX**. Note that the encoding procedure happens only over one of the qubits of the entangled pair—the one belonging to Alice—and hence the encoding operation only gets access that single qubit. Since the transformation is

applied in place on the qubit, there is no need for the operation to return anything
and its return type can be defined as Unit.

The decoding operation will accept both qubits as input parameters Decode
(qubit1 : Qubit, qubit2 : Qubit) and will return two classical bits as its
output. The operation corresponds to Bob's activities in the protocol and represents
his procedure of decoding of the two classical bits out of the involved qubits—the
reverse Bell circuit and two independent Z-basis measurements of each of the qubits.
Contrary to encoding, decoding requires access to both of the qubits. The operation
is defined in Listing 5.47.

```
operation Decode(qubit1 : Qubit, qubit2 : Qubit) : (Bool, Bool) {
    CNOT(qubit1, qubit2);
    H(qubit1);

    let result1 = IsResultOne(M(qubit1));
    let result2 = IsResultOne(M(qubit2));
    return (result1, result2);
}
```
Listing 5.47 Decoding operation of the superdense coding protocol.

With these two major building blocks in place, we are ready to define an operation
that will encapsulate the entire protocol. The operation RunDenseCoding (value1
: Bool, value2 : Bool) from Listing 5.48 accepts two input boolean parameters corresponding to the bits that will be encoded and, making internal use of the
encoding and decoding operation we just defined, executes the entire superdense
coding protocol. To achieve that, the operation allocates two qubits and entangles
them creating the initial system state $|\Phi^+\rangle$. Finally, it returns the two bits decoded by
Bob so that the orchestrating code can use them for verifying whether the protocol
was successful.

```
operation RunDenseCoding(value1 : Bool, value2 : Bool) : (Result,
    Result) {
    use qubits = Qubit[2];

    PrepareEntangledState([qubits[0]], [qubits[1]]);
    Encode(value1, value2, qubits[0]);

    return Decode(qubits[0], qubits[1]);
}
```
Listing 5.48 Single run of a superdense coding protocol in Q#.

To round things off, we introduce the orchestration code that will invoke
RunDenseCoding (value1 : Bool, value2 : Bool) an arbitrary number of
times and that would act as an entry point for our application. In previous
examples across this book, we settled on 4096 repetitions as the rule of thumb
for verifying our code. The code in Listing 5.49 provides all the necessary
orchestration capabilities and keeps track of how many times the encoded and
decoded messages agreed with each other. To randomize the input, we make

use of `DrawRandomBool` (`successProbability : Double`) operation from the
`Microsoft.Quantum.Random` namespace.

```
@EntryPoint()
operation Main() : Unit {
    mutable successCount = 0;
    let runs = 4096;
    for run in 1..runs {
        let (alice1, alice2) =
            (DrawRandomBool(0.5), DrawRandomBool(0.5));
        let (bob1, bob2) = RunDenseCoding(alice1, alice2);
        set successCount += alice1 == bob1 and alice2 == bob2
            ? 1 | 0;
    }

    Message("Success rate: "
        + DoubleAsStringWithFormat(100. * IntAsDouble(
            successCount) / IntAsDouble(runs), "N2"));
}
```

Listing 5.49 Superdense coding Q# orchestration code.

Equipped with such a Q# program, we can now execute it to verify the correctness
of the superdense coding protocol. Ideally, if no errors happen, we expect the encoded
input and the decoded output to always be the same. Under perfect conditions, this
is what we should see in the output of our program.

```
Success rate: 100.00
```

Listing 5.50 The expected output of the superdense coding test program.

5.7   Entanglement as a Resource

The entanglement-based communication protocols discussed here all share a very
important trait—they use entanglement as a transient resource, sort of a fuel, to be
able to execute. Viewing entanglement as a resource is a unique way of approaching
the entanglement phenomenon from the quantum informational theory perspective,
and the idea was first proposed by William Wootters [8]. He noted that as entanglement facilitates quantum information exchange, a stable established scheme and
infrastructure for *replenishing* it is required—a notion critical to further growth and
maturation of the field. Additionally, just like any other resource, such as energy or
information, entanglement needs to be quantifiable.

This is a very valuable perspective. Indeed, if quantum information theory, and
more specifically quantum communication, are to graduate to anything more than
a scientific curiosity and truly revolutionize our approach to communication and
information, it will most likely be because of the features facilitated by entanglement.
Commoditization of entanglement, as well as ensuring stability and coherence of
entangled particles, is thus one of the keys to unlocking and harnessing the true
potential buried in quantum hardware.

References

1. Bennett, C. H., & Wiesner, S. J. (1992). Communication via one- and two-particle operators on Einstein-Podolsky-Rosen states. *Physical Review Letters, 69*(Nov), 2881–2884.
2. Bennett, C. H., Brassard, G., Crépeau, C., Jozsa, R., Peres, A., & Wootters, W. K. (1993). Teleporting an unknown quantum state via dual classical and Einstein-Podolsky-Rosen channels. *Physical Review Letters, 70*(Mar), 1895–1899.
3. Boschi, D., Branca, S., De Martini, F., Hardy, L., & Popescu, S. (1998). Experimental realization of teleporting an unknown pure quantum state via dual classical and Einstein-Podolsky-Rosen channels. *Physical Review Letters, 80*(6), 1121–1125.
4. Bouwmeester, D., Pan, J.-W., Mattle, K., Eibl, M., Weinfurter, H., & Zeilinger, A. (1997). Experimental quantum teleportation. *Nature, 390*(6660), 575–579.
5. Bussieres, F., Clausen, C., Tiranov, A., Korzh, B., Verma, V. B., Nam, S. W., et al. (2014). Quantum teleportation from a telecom-wavelength photon to a solid-state quantum memory. *Nature Photonics, 8*(10), 775–778.
6. Cleve, R., Hoyer, P., Toner, B., & Watrous, J. (2004). Consequences and limits of nonlocal strategies. In *Proceedings. 19th IEEE Annual Conference on Computational Complexity* (pp. 236–249).
7. Collins, G. P. (1998). Quantum teleportation channels opened in Rome and Innsbruck. *Physics Today, 51*(2), 18–21.
8. Evans, J., & Thorndike, A. S. (Eds.). (2007). *Quantum mechanics at the crossroads. New perspectives from history, philosophy and physics.* Berlin: Springer.
9. Greenberger, D. M., Horne, M. A., & Zeilinger, A. (2007). Going beyond Bell's theorem.
10. Halder, M., Beveratos, A., Gisin, N., Scarani, V., Simon, C., & Zbinden, H. (2007). Entangling independent photons by time measurement. *Nature Physics, 3*(10), 692–695.
11. Mattle, K., Weinfurter, H., Kwiat, P. G., & Zeilinger, A. (1996). Dense coding in experimental quantum communication. *Physical Review Letters, 76*(Jun), 4656–4659.
12. Mermin, N. D. (2007). *Quantum computer science. An introduction.* Cambridge: Cambridge University Press.
13. Ren, J.-G., Xu, P., Yong, H.-L., Zhang, L., Liao, S.-K., Yin, J., et al. (2017). Ground-to-satellite quantum teleportation. *Nature, 549*(7670), 70–73.
14. Rieffel, E. G., & Polak, W. H. (2014). *Quantum computing: A gentle introduction.* Cambridge: MIT Press.
15. Sherson, J. F., Krauter, H., Olsson, R. K., Julsgaard, B., Hammerer, K., Cirac, I., et al. (2006). Quantum teleportation between light and matter. *Nature, 443*(7111), 557–560.
16. Tsirelson, B. (2021). Quantum information processing lecture notes. https://www.tau.ac.il/~tsirel/Courses/QuantInf/syllabus.html. Retrieved 06 July 2021.
17. Wang, C., Deng, F.-G., Li, Y.-S., Liu, X.-S., & Long, G. (2005). Quantum secure direct communication with high-dimension quantum superdense coding. *Physical Review A, 71*(03), 44305.

Chapter 6
Quantum Key Distribution

So far we have covered a wide array of topics around the nature of quantum computation–the theoretical aspects of qubits, superposition, single-qubit gates, multi-qubit gates, entanglement and several other interesting concepts from the area of quantum information theory and quantum communications.

In this chapter we will shift our attention to the field of quantum cryptography, and we will explore a few interesting key distribution protocols based on quantum phenomena. We shall also learn how to securely exchange a key between two parties using qubits transmitted over public channels and, naturally, realize the discussed protocols with basic implementations in Q#.

6.1 Key Distribution in Cryptography

While there is obviously a lot more to the field of cryptography than just key exchange, in principle, a securely shared symmetric key is enough to guarantee perfect protection against message cracking. This is thanks to a message encryption technique called *one-time pad*, which is information-theoretically secure, that is, it cannot be cracked because the *ciphertext* produced by it would not provide the attacker with any information about the original message.

The basic theory behind it is remarkably simple. Consider a 1-byte message that Alice would like to send over to Bob, as shown in the Table 6.1.

Of course in reality sending a single letter F is of questionable usefulness, but it makes for a good example. Because the message consists eight bits, the ASCII letter F that is being sent is one of the 2^8 possible values the message can take. In other words, it is one of the possible 256 manifestations of the message content. An adversary party, Eve, could guess the message with the probability of 1 in 256.

© The Author(s), under exclusive license to Springer Nature Switzerland AG 2022   181
F. Wojcieszyn, *Introduction to Quantum Computing with Q# and QDK*, Quantum Science and Technology, https://doi.org/10.1007/978-3-030-99379-5_6

Table 6.1   Uppercase letter F encoded in ASCII

message	0	1	0	0	0	1	1	0

Table 6.2   A random key

key	0	1	1	1	1	1	1	0

Table 6.3   A ciphertext created from *message* XOR *key*

ciphertext	0	0	1	1	1	0	0	0

Now let us imagine that Alice and Bob happen to shared a secret key, known only to them, that has the same length as the message—also eight bits (Table 6.2).

The key is represented here by a random bit sequence—01111110, which happens to be the tilde ~ character in ASCII, but that is irrelevant. It is, like the message, one of the possible 256 values for the key.

What Alice can do, is put the two together using XOR logic, to produce the ciphertext as shown in Table 6.3.

$$message \oplus key = ciphertext \tag{6.1}$$

The resulting ciphertext is 00111000, which is actually the character 8 in ASCII, though that is yet again inconsequential. The important thing is that the message and the key together produced another representation of the original message, yet again one of the 256 values in the 8-bit value pool. If the attacker, Eve, intercepts the ciphertext on its way from Alice to Bob, she would gain no information about the original message at all. All she has is one of the 256 8-bit values—which is as useful as the 8-bit value she could have guessed herself with the probability of 1 in 256.

Bob, on the other hand, can rely on another property of XOR, namely its reversibility. Applying XOR between the ciphertext and the key, would product the original message back. The reason why this works is because applying XOR between two copies of the same value, produces 0, so:

$$ciphertext \oplus key = (message \oplus key) \oplus key = message \tag{6.2}$$

This gives Alice and Bob a safe way of exchanging a message, even on a public channel. The prerequisites for the one-time pad technique to be unbreakable, are the following:

1. the key would be completely random
2. the key must be at least the length of the message to encrypt
3. the key must not be reused and naturally, must be kept secret by both parties (especially in transit).

Quantum key distribution protocols can provide guarantee to the parties engaging in cryptographic message exchange that the first and last requirements from the list above are going to be fulfilled. In particular the randomness achieved through the employment of quantum hardware would be the randomness underwritten by the laws of nature and particularly attractive for the cryptographic use cases. This is in stark contrast to classical computing, where most of the *software based* random number generators in various programming frameworks and libraries are not suitable for high sensitivity cryptographic scenarios. And with regards to ensuring that the key is kept safe even when exchanged over the public channel, quantum key distribution protocols, by taking advantage of the unique properties of quantum phenomena, provide a provable and reliable way of detecting an eavesdropper attempting to steal the key in transit between the two parties. This allows the engaged parties to abort the protocol if such intrusive behavior is determined, which dramatically improves the security of the key, as the two actors only have two focus on protecting the key once in their possession.[1]

Djordjevic [8] provides two main taxonomies for quantum key distribution schemes—one depending on the detection strategy used, and the other depending on the organization of quantum devices participating in the protocol.

In the first taxonomy, we can speak about two generic categories of QKD protocols:

1. **Discrete-Variable QKD**, typically built around single photon detectors, dealing with individual discrete events and thus measurement outcomes that form a finite set. The main security guarantee here are the no-cloning theorem and the indistinguishability of non-orthogonal states.
2. **Continuous-Variable QKD**, which rely on measurements of the quadratures of the electric field of the incident light, yielding a continuum of measurement outcomes. Security-wise, relies on uncertainty principle.

Secondly, we can also categorize quantum key distribution protocol into three of the following subgroups:

1. **Device-dependent QKD**. In these types of key exchange schemes, quantum source is placed with one of the actors, typically Alice, while the quantum detector is located at Bob's side.
2. **Source-device-independent QKD**. In these schemes, the quantum source is placed with a third party (Charlie), while the detectors are located with both Alice and Bob
3. **Measurement-device-independent QKD**. In such setup, the quantum sources are located with both Alice and Bob, but the detectors are placed with Charlie, who performs partial Bell measurements

---

[1] It is outside of the scope of this book to discuss cryptographic techniques any further, however a great introduction to one-time pad and other algorithms can be found in [17], which dedicates a very digestible chapter to this "simplest and most secure way to keep secrets".

In this study we shall focus on *discrete-variable QKD* only, and protocol descriptions according to *device-dependent QKD* (BB84, B92) and *source-device-independent QKD* (E91) only. While that does not correspond to the most practical and efficient set of schemes from the commercial application perspective, the protocols we will cover here form a solid basis for introduction to quantum key distribution.

Even though quantum-powered distribution of encryption keys may seem like science fiction, it has not only been realized in laboratory, but is already employed in various quantum communications networks. Among the most impressive examples has been the SwissQuantum QKD network, in operation between 2009 and 2011, which demonstrated that QKD was mature enough to be deployed and integrated into telecommunications networks and "has proven its reliability and robustness in a real life environment outside of the laboratory." [16].

Another terrific example is the Chinese research project "The Quantum Experiments at Space Scale" (QUESS), which launched the low-Earth-orbit Quantum Space Satellite nicknamed *Micius* in 2016 [10], or the recent work in the area of underwater quantum key distribution [18].

Within the QUESS project, in 2018, a group led by Sheng-Kai Liao reported a staggering success [12], performing quantum key distribution between the satellite and ground stations located in Xinglong and Graz, on other sides of the globe, with a separation of 7600 km. The keys were then used for both one-time pad encryption to transmit images, as well as to encrypt a video conference call. The researchers stated that this can be viewed as "a simple prototype for a global quantum communications network."

The technique used by Chinese researchers was the BB84 quantum key distribution protocol, which will be our focus point next.

6.2   BB84 Protocol

The BB84 quantum key distribution protocol[2] was developed by Charles Bennett from IBM Research and Gilles Brassard from the University of Montreal in 1984 [4], who were the first ones to propose to use the principles of quantum information theory, as well as certain constraints of quantum mechanics, in the field of cryptography.

Bennett and Brassard sparked the birth of an entire new branch of quantum research—quantum cryptography. Through their work they also laid foundations for various other quantum key distribution protocol ideas that followed. The specific variant designed Bennett and Brassard has since been revised multiple times and has become an important building block for quantum information teaching curriculums. It is also relatively simple to follow, and hence is a suitable place for us to start the journey into the quantum key distribution world.

---

[2] The name comes from the year the paper was published and the first letters of the authors' last names—a relatively common practice for landmark papers.

Currently, our best classical cryptographic protocols are designed around problems that are deemed too hard to solve, such as finding large prime factors. But in the case of BB84 and related protocols, the security is not based on an idea that certain class of problems is difficult, but is instead underpinned by the physical laws of nature. The ingenuity of the BB84 protocol lies in its remarkable simplicity. One only needs to be familiar with two fundamental quantum mechanical concepts to be able to grasp the idea behind it.

First of all, the protocol relies on indistinguishability of non-orthogonal states. When we first introduced the concept of superposition in Sect. 2.2, we mentioned that it is, in fact, basis dependent, and this is the phenomenon explored by BB84. Given two of the so-called unbiased bases, such as the computational basis (Z-basis) $\{|0\rangle, |1\rangle\}$ the X-basis $\{|+\rangle, |-\rangle\}$, when the system is prepared in one of the orthogonal states of one of the bases, all possible measurement outcomes with respect to the other basis are equally likely. In other words—a definite basis state in one of these bases, is the perfect uniform superposition in the other basis. Using this principle, BB84 authors proposed to encode information, not as usually, in two orthogonal states making up one basis, but by randomly switching between two unbiased bases. As a consequence of this, the information can only be retrieved from the quantum system by correctly applying the basis used in the encoding process. Additionally, such approach to handling of information prevents even partial access to information by unauthorized parties.

Secondly, the protocol explores the fact that according to the no-cloning theorem, which we covered in Sect. 2.7, an unknown quantum state cannot be reliably copied. While eigenstates of a known basis can be cloned without any limitations, the BB84 protocol, by employing two mutually unbiased bases to encode information, benefits from the restrictions of no-cloning theorem with quite profound consequences.

BB84 additionally incorporates mechanisms for detecting possible eavesdroppers on the communication channel. This in turns means that BB84 can be theoretically used to securely exchange a symmetric encryption key over a public *unsecured* channel and without a need for sharing any prior information between the parties.

In order to best be able to follow the steps of the BB84 protocol, let us set the stage on which it can play out. Alice and Bob would like to communicate between each other privately, and thus would like to share an encryption key. To get going, they establish that Alice will send a truly random key to Bob, which they can later use to engage in symmetric cryptography. Unbeknownst to them, the evil Eve is plotting against Alice and Bob and will attempt to hijack the key in transit.

The key will have a length of n classical bits. According to BB84, Alice will use two distinct bases, Z-basis and X-basis, to encode those bits into qubits—switching between them randomly. In other words, she shall use one of the following four quantum states denoted by $|\psi_n\rangle$ to send a classical bit value encoded into a qubit to Bob.

$$|\psi_1\rangle = |0\rangle = 0 \qquad |\psi_3\rangle = |+\rangle = \frac{1}{\sqrt{2}}\left(|0\rangle + |1\rangle\right) = 0 \qquad (6.3)$$

$$|\psi_2\rangle = |1\rangle = 1 \qquad |\psi_4\rangle = |-\rangle = \frac{1}{\sqrt{2}}\left(|0\rangle - |1\rangle\right) = 1$$

As outlined in (6.3), in the BB84 scheme classical 0 and 1 have two representations each. $|\psi_1\rangle$ and $|\psi_3\rangle$ can be used to represent classical bit value 0, while $|\psi_2\rangle$ and $|\psi_4\rangle$ can be used to represent classical 1. Notice that with such configuration, we are able to encode the same classical bit value using two non-orthogonal states, one being an eigenstate of the Z-basis, the other an eigenstate of the X-basis.

BB84 also prescribes that for a key of a desired length n, $4n$ random bits should be generated and transported between the parties. As we shall see soon, this is required because of the efficiency of the protocol and because certain bits shall later be discarded for the purpose of detecting the eavesdropper. Since a single qubit is required for each bit that needs to be encoded, once her encoding process finishes, Alice possesses $4n$ qubits containing her original bit values encoded using two different bases.

In the next step, using a quantum communication channel, Alice sends the qubits to Bob, who then proceeds to measure them. This communication can happen over an unsecured public network, which is one of the strengths and advantages of BB84. Bob is unaware of how the data was encoded into the qubits, and therefore is forced to guess the bases that Alice used. Therefore, upon measurement, he randomly changes between the Z and X-basis measurements. Statistically speaking, Bob will choose the correct basis half of the time. Of course we know that when the qubit is in an eigenstate of Z-basis, its measurement in X-basis is maximally uncertain, and vice versa. A consequence of this is that if Bob guesses the basis right, then his measurement result will exactly agree with the value that Alice encoded, but if he chooses wrongly, because of a uniform superposition, he would get the correct classical bit result only for 50% of his measurements.

At this point Alice and Bob both possess arrays of classical bits—Alice the one she initially created randomly to encode into the qubits, while Bob the one that he read out from the received qubits. The length of those arrays is going to be the same, however, at indices where they used a different basis to encode/decode bit into the corresponding qubit bit, the values may differ. Because of that, the next thing they need to do, is to compare the bases they chose for each individual qubit—Alice to encode, and Bob to decode the information. They can do this over the classical information channel, such as phone, email or any other suitable medium. Obviously they should only announce the bases chosen—not the classical bit values, as those need to remain private. Alice and Bob would then discard all of the indices for which they used different bases and leave only the ones for which the applied basis was identical. Given that, statistically speaking, Bob would guess the correct basis half of the time, they should at this point be left with about $2n$ bits. We will not consider this here, as we are assuming ideal conditions for simplicity, but Alice and Bob might also establish a certain level of tolerance for deviation from the mean—for example, if the amount of bits left is significantly lower than $2n$, they may opt to restart the protocol from the beginning.

The reason why Alice and Bob still need to have approximately $2n$ of bits each, despite the fact they are only wanted to share a key of length n length, is that the remaining half—so n bits—will be *sacrificed* in order to perform an eavesdropper check. The way it plays out is as follows. Alice should randomly select n of the remaining $\approx 2n$ bits, and share their indices with Bob. This, again, can happen over a public channel, and allows Alice and Bob to compare their bit values (not bases anymore, but the actual values!). Since they already ensured that the bases their chose were identical for those specific bits, the values must be identical as well—and therefore they expect a 100% agreement. This is indeed what they will confirm when no eavesdropper tapped into their quantum communication channel. On the other hand, when they find out any mismatch, it suggests malicious activity—due to the very specific wave function collapse effect an eavesdropper would have on the qubits in transit.[3]

It is reasonable to ask at this point, how would an attacker's involvement look like. Consider what happens to the qubits when an attacker, Eve, tries to intercept the qubit in transit. The most optimal scenario for a threat actor would be to copy the qubit into her own storage and let the original continue over the network towards the intended recipient. However, as we already know, due to the no-cloning theorem, this is physically impossible to realize. As a result, the only attack Eve can undertake is to attempt to measure the qubit in transit, save the resulting classical bit and send the qubit back onto its original path to Bob. The obvious problem here is that a measurement irreversibly changes the state of the qubit, which would not matter if the qubit was in an eigenstate of a specific basis, and then was measured in that basis, but would definitely matter if it was measured in the wrong basis. In principle, just like Bob, Eve does not know in which basis to measure so she would need to guess too, and we already established that this would lead to only 50% success probability—and this is the core idea behind BB84. In those 50% of cases when Eve guesses the bases wrongly, she causes the state vector to collapse to one of the eigenstates of her wrong basis. This naturally means that Bob, even though he is using the appropriate basis to decode the qubit, is now measuring against a superposition state, and is equally likely to read out a 0 or 1. As a consequence, in 50% of those cases where an attacker guess wrongly, Bob's decoded classical bit would differ from Alice's. As a result, Alice and Bob would see a discrepancy of about 25% in the n set they used for eavesdropper detection.

What BB84 prescribes is that if Bob and Alice determine that their bit values disagree even though their bases agree, they should abort the protocol as it indicates that they might have fallen victims to an attacker.

In the situation when no eavesdropper was detected, Alice and Bob proceed to throwing away the n bits they used for the security check and can safely use the remaining n bits as their intended shared encryption key. At the end the price they paid for the theoretical security of the protocol and for the ability to sniff out a potential eavesdropper is that they needed to transfer four times the amount of required qubits between them—$4n$, instead of n, leading to a 25% BB84 protocol efficiency rate.

The overview of the protocol, excluding the eavesdropper check, is shown in Table 6.4. Only the bits with probability of correct decoding equal to 1 can be kept.

---

[3] Yet again, for simplicity we ignore here any potential quantum communication errors, noise, packet loss and consider ideal conditions only.

Table 6.4 Overview of the BB84 protocol

Alice bit	$a = 0$				$a = 1$			
Alice encoding basis	Z		X		Z		X	
Alice state sent	$|0\rangle$		$|+\rangle$		$|1\rangle$		$|-\rangle$	
Bob decoding basis	Z	X	Z	X	Z	X	Z	X
Probability of correct decoding	1	0.5	0.5	1	1	0.5	0.5	1

The original description of BB84 was premised on switching the encoding of information between Z $\{|0\rangle, |1\rangle\}$ and X $\{|+\rangle, |-\rangle\}$ bases, however, in principle, any two mutually unbiased bases would be applicable too.

Despite its obvious advantages and the theoretical security, the biggest obstacles for the BB884 protocol are the well-known problems of cheaply and reliably transferring quantum objects between different network nodes. The quantum communications infrastructure is currently limited, expensive, error-prone, and still far from mainstream usage. Large distance successful realizations of quantum key distribution protocols remain challenging, in particular due to fibre attenuations, environment noise, challenges related to scaling and limitations of repeater technology. At the moment the longest successful BB84 key exchange using optical fiber telecommunication network infrastructure has been performed over the distance of 421km [5].

The original basic BB84 variant is very difficult to scale because it was based on a single photon source, which creates a number of challenges. Instead, faint lasers are often used in place, but at the expense of increasing the attack-surface. In principle, such multi-photon signals are susceptible to the so-called *photon-number splitting* attacks, allowing Eve to send over beam-splitted photon signals to Bob, keeping the other photon, now entangled with the one travelling to Bob, to herself, effectively bypassing the restrictions of the no-cloning theorem [13].

In 2004, a Swiss research team from the University of Geneva developed a BB84 spin-off, which become known, after the names of the team of creators, as the *SARG04 protocol* [15]. It was specifically designed to address the Achilles heel of BB84, the photon-number splitting attacks, and was intended to be its replacement. In the servers of the SwissQuantum QKD network, both BB84 and SARG04 were supported, though the ultimate production operation was done on top of the SARG04 protocol only.

As we just learnt, the theory behind BB84 is not very difficult to follow, once you understand the basics behind the qubit measurement in different bases. Similarly, we can relatively easily built an illustrative example of BB84 in Q# code. We will need to improvise a little and take certain shortcuts—after all the whole protocol will have to play itself out in a single process—but nevertheless, we shall be able to reflect all the core concepts and ideas behind BB84 in a small Q# program.

To keep things manageable, we shall split the entire protocol into a number of smaller Q# operations and functions. The main functionality will be encapsulated in a RunBB84Protocol (roundtrips : Int, rountripSize : Int, eavesdropperProbability : Double) operation, which will orchestrate execution of all these other callables following the protocol steps we just outlined theoretically. The orchestration operation is shown in its entirety in Listing 6.1.

```
operation RunBB84Protocol(roundtrips : Int, rountripSize : Int,
    eavesdropperProbability : Double) : Unit {
    mutable aliceKey = [false, size = 0];
    mutable bobKey = [false, size = 0];

    for roundtrip in 0..roundtrips-1 {
        use qubits = Qubit[rountripSize];

        // 1. Alice chooses her bases
        let aliceBases = GenerateRandomBitArray(rountripSize);

        // 2. Alice chooses her random bits
        let aliceValues = GenerateRandomBitArray(rountripSize);

        // 3. Alice encodes the values in the qubits using the
            random bases
        EncodeQubits(aliceValues, aliceBases, qubits);

        // 4. Eve attempts to evesdrop based on the configurable
            probability
        Eavesdrop(eavesdropperProbability, qubits);

        // 5. Bob chooses his bases
        let bobBases = GenerateRandomBitArray(rountripSize);

        // 6. Bob measures qubits using the random bases
        let bobResults = DecodeQubits(bobBases, qubits);

        // 7. Alice and Bob compare their bases and throw away
            those values that were encoded/decode in non-matching
            bases
        let (aliceRoundTripResult, bobRoundTripResult)
            = CompareBases(aliceBases, aliceValues, bobBases,
                bobResults);

        // 8. Append both key from this roundtrip to the overall
            key
        // repeat however many times needed
        set aliceKey += aliceRoundTripResult;
        set bobKey += bobRoundTripResult;
    }

    // 9. Perform the eavesdropper check
    let (errorRate, trimmedAliceKey, trimmedBobKey) =
        EavesdropperCheck(aliceKey, bobKey);

    // 10. Output the resulting keys and additional useful info
    ProcessResults(errorRate, trimmedAliceKey, trimmedBobKey,
        eavesdropperProbability);
}
```

Listing 6.1 Skeleton of the BB84 implementation in Q#.

The operation takes three input parameters, that will allow us to fine-tune its execution and test the behavior of the protocol under various conditions. The first two are the roundtrips and roundtripSize, which will help us control the interactions between Alice and Bob. For example, if we aim to execute the BB84 for an $n = 128$ bit key, it would mean that we would need to allocate at minimum $4n = 512$ qubits—way beyond the capabilities of current quantum computing hardware or, especially, simulators. Instead, we can run the protocol in a number of roundtrips—using roundtripSize qubit batches in each roundtrip. A realistic approach for executing this code on a quantum simulator would be to keep the roundtripSize under 20.

In addition to all of that, the eavesdropper probability can be passed into the entry operation from Listing 6.1 too. It is represented by a double—to express percentage values—and will let us invoke the code with eavesdropper participation determined by arbitrary probabilities as well.

Before the procedure begins, we shall allocate two mutable arrays, one to hold the key of Alice and the other of Bob. The arrays must be mutable, as we shall continuously expand them after every roundtrip—given the roundtrip size limitations, each roundtrip will extend the key by a relatively small number of bits only. Alice's array aliceKey will be derived from her initial random bit set (source bits) while Bob's array bobKey will be based on his measurement results (result bits).

The roundtrips are simulated with a regular for loop, and each iteration allocates roundtripSize number of qubits. The first two steps of the protocol, as annotated in Listing 6.1 as well, are for Alice to generate her random bits and choose her random bases. Since both of these operations are effectively just generations of random bit arrays, they can be handled by the same helper operation, GenerateRandomBitArray (length : Int), which is shown in Listing 6.2.

```
operation GenerateRandomBitArray(length : Int) : Bool[] {
    mutable bits = [false, size = length];

    for i in 0..length-1 {
        let randomBit = DrawRandomBool(0.5);
        set bits w/= i <- randomBit;
    }

    return bits;
}
```

Listing 6.2 Q# operation to generate a random bit array.

The randomness of the bits in the newly created arrays is guaranteed by the DrawRandomBool (successProbability : Double) operation from the Microsoft.Quantum.Random namespace—the same one we used already several times across this book.

The next step, denoted in Listing 6.1 with number three, is for Alice to encode the randomly selected bits aliceValues into the qubits, using the randomly selected bases aliceBases as reference. This is done via the EncodeQubits (bits : Bool[], bases : Bool[], qubits : Qubit[]) operation.

```
operation EncodeQubits(bits : Bool[], bases : Bool[], qubits :
    Qubit[]) : Unit {
    for i in 0..Length(qubits)-1 {
        let valueSelected = bits[i];
        if (valueSelected) { X(qubits[i]); }

        // 0 will represent |0> and |1> computational (PauliZ)
            basis
        // 1 will represent |-> and |+> Hadamard (PauliX) basis
        let basisSelected = bases[i];
        if (basisSelected) { H(qubits[i]); }
    }
}
```

Listing 6.3 Encoding of qubits step of the BB84 protocol in Q#.

In this encoding operation, Alice will iterate through all the qubits, and, since the randomly generated arrays of bit values and bit bases have the same amount of elements, she will access the corresponding random value and random basis using the indexer. If the corresponding bit value was intended to be 1, she will apply the **X** gate to the qubit to perform the bit-flip operation. She will also adopt a rule that if the random basis bit is 0, she will use computational basis $\{|0\rangle, |1\rangle\}$ to encode the data in the qubit, while if it is 1, she will use the X-basis $\{|+\rangle, |-\rangle\}$. The practical consequence of such a scheme, is that in the latter case she would need to additionally apply the **H** unitary transformation to the qubit.

At this point we can—in our imagination at least—assume that the qubits fly over to Bob using a quantum communication link. Thus, in the next step, marked in Listing 6.1 with number four, we will introduce Eve, who will attempt to intercept and read those qubits. She does it with the operation Eavesdrop (eavesdropperProbability : Double, qubits : Qubit[]).

```
operation Eavesdrop(eavesdropperProbability : Double, qubits :
    Qubit[]) : Unit {
    for qubit in qubits {
        let shouldEavesdrop = DrawRandomBool(
            eavesdropperProbability);
        if (shouldEavesdrop) {
            let eveBasisSelected = DrawRandomBool(0.5);
            let eveResult = Measure([eveBasisSelected ? PauliX |
                PauliZ], [qubit]);
        }
    }
}
```

Listing 6.4 Eavesdropper involvement as part of BB84 implementation in Q#.

Eve gets a chance to interact with each of the qubits in the roundtrip batch, based on the eavesdropperProbability. If we have determined that eavesdropper can act on a given qubit, she shall randomly switch between the Z and X-bases, because she has to guess the bases, and can proceed to the measurement. The same DrawRandomBool (successProbability : Double) operation as used earlier, can act here as a reliable decision maker. Toggling between measurement bases in

Q# is very easily achieved using the Measure (bases : Pauli[], qubits :
Qubit[]) which we first introduced already back in Sect. 4.1. After the measurement, Eve obtains her result and the qubit then continues on its way to Bob.

The next step of the protocol foresees Bob receiving the batch of qubits. In principle, he shall follow the exact same procedure as Eve had to go through—he will need
to select a random basis for each qubit and measure it. In this implementation, annotated with step five in Listing 6.1, Bob will select all his random bases upfront using
GenerateRandomBitArray(length : Int) operation from Listing 6.2, just as
Alice did. Those bases are then captured into the bobBases variable.

Bob will then measure the qubits using his randomly selected bobBases as measurement guide—and this is expressed by the DecodeQubits(bases : Bool[],
qubits : Qubit[]) operation and marked in Listing 6.1 with step six.

```
operation DecodeQubits(bases : Bool[], qubits : Qubit[])
 : Bool[] {
    mutable bits = [false, size = 0];

    for i in 0..Length(qubits)-1 {
        let result = Measure([bases[i] ? PauliX | PauliZ], [
            qubits[i]]);
        set bits += [ResultAsBool(result)];
    }

    return bits;
}
```

Listing 6.5 Decoding of qubits step of the BB84 protocol in Q#.

In order to properly decode the qubits that arrived, Bob allocates a mutable array
to hold the bit results, iterates through the qubits, and measures each qubit randomly
choosing either the X or Z-basis. The measurement result is converted into a classical
bit using the ResultAsBool (input : Result) conversion function, from the
Microsoft.Quantum.Convert namespace.

At this point we can quickly regroup, to make sure we know exactly where
we are in BB84. Alice has sent her random bits, encoded into qubits and stored
them in aliceValues array. Her, also random, bases to encode those values
have been stored in aliceBases array. Bob's measurement results are located in
bobResults array, while his randomly chosen measurement bases are stored in
bobBases array. Eve interfered with the communication with the probability equal
to eavesdropperProbability though her measurement results and bases are not
stored anywhere as they are irrelevant for the protocol—what is important is whether
Alice and Bob will be able to detect her malicious presence at all.

The next step, marked with seven in Listing 6.1 is the first critical junction for the
BB84 protocol. Alice and Bob shall compare their bases at this point, and discard
the classical bits corresponding to the qubits that they (respectively) encoded and
decoded in non-matching bases. In other words, if the bases for a given qubit agree
for them, then they want to keep the corresponding source bit value (Alice) or the
corresponding measurement bit result (Bob). Based on sheer probabilities, given
that Bob was guessing the basis from two possible options, Alice and Bob should

have about 50% of original qubits involved in the roundtrip left. The deviations in the Q# implementation can be large though, because as explained in the beginning, due to hardware and simulator limitations, our rountripSize will be a relatively low number. The comparison function CompareBases (basesAlice : Bool[], bitsAlice : Bool[], basesBob : Bool[], bitsBob : Bool[]) is shown in Listing 6.6.

```
function CompareBases(length : Int, basesAlice : Bool[],
    bitsAlice : Bool[], basesBob : Bool[], bitsBob : Bool[]) : (
    Bool[], Bool[]) {
    mutable trimmedAliceBits = [false, size = 0];
    mutable trimmedBobBits = [false, size = 0];

    // compare bases and pick shared results
    for i in 0..length-1 {

        // if Alice and Bob used the same basis
        // they can use the corresponding bit
        if (basesAlice[i] == basesBob[i]) {
            set trimmedAliceBits += [bitsAlice[i]];
            set trimmedBobBits += [bitsBob[i]];
        }
    }

    return (trimmedAliceBits, trimmedBobBits);
}
```

Listing 6.6 Comparison of Alice and Bob bases step of the BB84 protocol in Q#.

The function will trim the bit arrays held by Alice and Bob to only keep the indices for which the corresponding basis choice was identical. Those are then returned from the function for processing in the main branch of the code as a tuple. There, they are reconstructed into two variables: aliceRoundTripResult and bobRoundTripResult. If there was no eavesdropper, those two arrays are expected to be identical.

Step eight in Listing 6.1 is straightforward—append the results of the single roundtrip for both Alice and for Bob to their respective mutable key arrays— aliceKey and bobKey. Based on roundtrip configuration that was passed into the orchestration operation, the procedure outlined in steps one to eight may repeat several times, to let the key grow to larger sizes. We purposely left eavesdropper check for later—to perform it once only, on a larger key.

Sniffing out a potential attacker on the communication channel is the last logical step in the protocol—and step nine in Listing 6.1, where it is also represented by an invocation of an EavesdropperCheck (aliceKey : Bool[], bobKey : Bool[]) operation.

```
operation EavesdropperCheck(aliceKey : Bool[], bobKey : Bool[]) :
    (Double, Bool[], Bool[]) {
    mutable trimmedAliceKey = [false, size = 0];
    mutable trimmedBobKey = [false, size = 0];
    mutable eavesdropperIndices = [0, size = 0];
```

```
for i in 0..Length(aliceKey)-1 {
    let applyCheck = DrawRandomBool(0.5);
    if (applyCheck) {
        set eavesdropperIndices += [i];
    } else {
        set trimmedAliceKey += [aliceKey[i]];
        set trimmedBobKey += [bobKey[i]];
    }
}
mutable differences = 0;
for i in eavesdropperIndices {
    // if Alice and Bob get different result, but used same
       basis
    // it means that there must have been an eavesdropper (
       assuming perfect communication)
    if (aliceKey[i] != bobKey[i]) {
        set differences += 1;
    }
}
let errorRate = IntAsDouble(differences)/IntAsDouble(Length(
    eavesdropperIndices));
return (errorRate, trimmedAliceKey, trimmedBobKey);
}
```

Listing 6.7 Eavesdropper check step of the BB84 protocol in Q#.

To facilitate the check for eavesdropper presence, Alice and Bob must now randomly choose 50% of their remaining classical bits and sacrifice them by performing public comparison. This can be expressed in Q# in a number of ways, but in our case, we shall simply iterate through all the bits that Alice/Bob now possess and, using our good friend DrawRandomBool(successProbability : Double), randomly deciding whether a given bit should be used for eavesdropper check, or whether it should be kept as part of the expected final key. If it should be kept, we add it to two helper mutable arrays, trimmedAliceKey and trimmedBobKey, where the ultimate keys will be located. On the other hand, if we determine that a bit ought to be used for eavesdropper detection, we need to note down its index in yet another mutable helper array, eavesdropperIndices.

Once the comparison bit indices are known and stored in eavesdropper Indices, all that is left is to loop through that array and pick the corresponding bit values out of Alice's and Bob's original key arrays for comparison. Leaving quantum errors aside for simplicity of this discussion, if at this point there is any difference between Bob and Alice, it means that there was an eavesdropper on their quantum communication channel, as it should not be possible for them to possess different classical bit value if they chose the same basis. Of course as we already extensively discussed, Eve as the attacker could also guess the basis correctly and if all three of them agreed on a basis, then no difference would be recorded. But Eve will be wrong about 50% of time, which will then cause *some* discrepancies between Alice and Bob, and this is exactly what they are after.

Using the running total of recorded differences and the total of bits selected for eavesdropping check, Alice and Bob can compute the error rate, which they can use as a confidence indicator that no one intercepted their communication—captured as a `double` in the `errorRate` variable. In this idealistic example, we shall use 0.0 as the error threshold—if that level is exceeded, meaning there was any difference between the Alice's and Bob's bit values in the sampled set, the protocol will be aborted. The operation then returns this error rate, as well as the trimmed keys.

The last, tenth step of Listing 6.1 is a helper function shown in Listing 6.8, which does not contribute any protocol logic anymore, and exists merely for our convenience to summarize the outcome of the entire process and provide some useful output information.

```
function ProcessResults(errorRate : Double, aliceBits : Bool[],
    bobBits : Bool[],eavesdropperProbability : Double) : Unit {
    Message($"Alice's key: {BoolArrayAsBigInt(aliceBits)}");
    Message($"Bob's key:   {BoolArrayAsBigInt(bobBits)}");

    Message($"Error rate: {errorRate * IntAsDouble(100)}%");
    if (errorRate > 0.0) {
        Message($"Eavesdropper detected!");
    }

    if (EqualA(EqualB, aliceBits, bobBits)) {
        Message($"Running the protocol with eavesdropping
            probability {eavesdropperProbability} was successful.
            ");
    } else {
        Message($"Running the protocol with eavesdropping
            probability {eavesdropperProbability} was
            unsuccessful.");
    }
    Message("");
}
```

Listing 6.8 Helper function to display BB84 protocol results.

This final piece is a formality—we use it to print out the final keys, which after the trimming operation, caused by the eavesdropper check, are going to be (approximately) halved again. Due to the statistical nature of the protocol, subsequent runs may yield keys of different results. This is not a problem per se, though it is something we do not account for in our Q# implementation. It is, however, possible to imagine an implementation that would continue to add extra roundtrips, until a specific key size is reached. At this point, of course, the keys are expected to be identical for both Alice and Bob, and ready to be used to engage in cryptographic activities. As such ,the keys naturally must not be revealed to anyone, nor can they be compared anymore. However, what the helper function from Listing 6.8 does, is it additionally performs an explicit comparison of both keys—this is for our convenience to verify the correctness of all the steps our Q# code executed. This comparison is facilitated by a built-in equal-

ity check function EqualA<'T> (equal : (('T, 'T) -> Bool), array1 :
'T[], array2 : 'T[]) from the Microsoft.Quantum.Arrays namespace.

Q# does not have a standard library functionality for serializing a boolean array
into a bit string, however printing a long series of zeroes and ones making up the
shared key would not be particularly readable anyway. Instead, we can use the
BoolArrayAsBigInt (a : Bool[]) function which can convert a bit array into
a corresponding big integer, which would be helpful for illustrative purposes.

The final piece left is to invoke the protocol—we do it by performing a call into
the main orchestration operation from Listing 6.1.

```
@EntryPoint()
operation Start() : Unit {
    RunBB84Protocol(32, 16, 0.0);
}
```

Listing 6.9 Execution of the BB84 protocol in Q# without any configured eavesdropper.

The first trial run uses 32 roundtrips of batch size 16, which amounts for a total
of 512 qubits involved in the protocol. After halving the bits twice—once for basis
comparison, and once for eavesdropper check—we should expect a key of about 128
bit length to be remaining. In this particular run, the eavesdropper probability is set
to 0%, thus we expect the protocol to be successfully executed.

An example of a successful output should look as that from Listing 6.10.

```
Alice's key: 49422817911153173607950404666615314050946
Bob's key:   49422817911153173607950404666615314050946
Error rate: 0%
Running the protocol with eavesdropping probability 0 was
    successful.
```

Listing 6.10 Sample output from running the code from Listing 6.9.

Once we activate the eavesdropper, with engagement probability 1, we should see
the failure of the protocol.

```
@EntryPoint()
operation Start() : Unit {
    RunBB84Protocol(32, 16, 1.0);
}
```

Listing 6.11 Execution of the BB84 protocol in Q# with a certain eavesdropper participation.

The output would now resemble the one from Listing 6.12, obviously it will differ
a little bit from run to run due to probabilistic nature of the protocol.

```
Alice's key: 1773265090803148995264498418075150 5962
Bob's key:   1610397341932748960902036460653380 3818
Error rate: 25.517241379310345%
Eavesdropper detected!
Running the protocol with eavesdropping probability 1 was
    unsuccessful.
```

Listing 6.12 Sample output from running the code from Listing 6.11.

This is aligned with our expectations—as discussed earlier, eavesdropper presence on each qubit should create about 25% error rate and cause Alice and Bob to detect its intrusive presence and be able to abort the protocol.

Finally, if we reduce the eavesdropper engagement to 50%, we should expect the error rate to go down to about 12.5%.

```
@EntryPoint()
operation Start() : Unit {
    RunBB84Protocol(32, 16, 0.5);
}
```

Listing 6.13 Execution of the BB84 protocol in Q# with eavesdropper probability set to 50%.

This should indeed be possible to be observed in the result, and at this point it would be up to Alice and Bob to decide if they are confident enough in such result, or should they restart the protocol from scratch.

```
Alice's key: 50351807120299654064867298338958320340
Bob's key:   50345348489839854975502144251925649000
Error rate: 13.432835820895523%
Eavesdropper detected!
Running the protocol with eavesdropping probability 0.5 was
    unsuccessful.
```

Listing 6.14 Sample output from running the code from Listing 6.13.

6.3 B92 Protocol

Eight years after the BB84 quantum key distribution protocol was published, Charles Bennett proposed its alternative variant [3], which later became known as the B92 protocol for quantum key distribution.

Bennett realized that a measurement that "fails to disturb each of two non-orthogonal states also fails to yield any information distinguishing them". As a consequence, in the B92 scheme, he simplified the original BB84 approach by employing two non-orthogonal states only, making up a single non-orthogonal basis, instead of four quantum states forming two mutually unbiased bases. From that perspective, B92 protocol really highlights how quantum key distribution protocols can guarantee information security through the indistinguishability of non-orthogonal state [14].

The non-orthogonal basis proposed in B92 to encode a classical bit value x was made up of states $|0\rangle$ and $|+\rangle$ from the Z and X-bases respectively is shown in (6.4).

$$|\psi\rangle = \begin{cases} |0\rangle & \text{when } x = 0 \\ |+\rangle = \frac{1}{\sqrt{2}}(|0\rangle + |1\rangle) & \text{when } x = 1 \end{cases} \tag{6.4}$$

Beyond this difference, BB84 and B92 are effectively the same protocol, and from the physical realization perspective are interchangeable under the same physical conditions [11]. Just like in the case of BB84, the usage of states originating from

Z and X-bases is not mandatory, the protocol can work equally well using a pair of non-orthogonal eigenstates from any two mutually unbiased bases.

The protocol steps would largely follow the BB84 steps outlined in Sect. 6.2. It starts with Alice randomly generating an array of classical bits that will be used as the basis for the shared key. Contrary to BB84, she no longer needs to keep track of the bases she chooses, because she shall encode the random bits using the methodology defined in Eq. 6.4—if she generates a random 0, she transforms the qubit into state $|0\rangle$, and when she generates 1, the qubit will need to be transformed into $|+\rangle$. The former can be achieved by simply not applying any gates to the qubit (or using the no-op $\mathbf{I}$ gate), the latter by applying the $\mathbf{H}$ transformation.

As we will soon find out, the main disadvantage of B92 is that it is actually only half as efficient as BB84, and thus to transfer a key of length n bits, Alice must initially generate $8n$ bits—and she will naturally require as many qubits to encode them.

Once generated, Alice sends over her qubits, now containing all the encoded information, to Bob. While Alice operated on a single non-orthogonal basis only, a hybrid Z and X-basis, Bob can only measure the qubits using orthonormal bases. Because of that, he will perform the measurements the same way as he would have done it in BB84, namely by randomly switching between the Z and X-bases—after all he does not know how Alice encoded the data, so the best he can do is guess. Just like in BB84, Bob will pick the correct basis roughly 50% of the time—and he proceeds to measure the qubits and note down all the measurement results.

Each of the qubits in this scheme, can be described by three classical bits. The first one, we shall refer to it as a, is the value that Alice encoded into the qubit. The second one, and we will call it b, represents the basis that Bob randomly selected for measurement. Imagine that when he needs to choose the basis for measurement Bob generates a random classical bit (or flips the proverbial coin)—which is really the value of b. If $b = 0$, he measures the qubit in the Z-basis, and if $b = 1$, he performs the observation in the X-basis. Upon measurement, irrespective of the basis chosen, Bob obtains a classical measurement result, which is the third classical bit we need to be aware of—we will call it r.

Let us recap the scheme up to this point quickly. Alice generated a classical bit a, encoded it into a qubit using the non-orthogonal basis which was a mix of Z and X-bases and sent it over to Bob. Bob switches between the Z and X observables for the measurement process based on the value of his randomly generated bit b and obtains a classical result r. This procedure would be repeated $8n$ times for a key of length n.

The next step in BB84 would be the basis comparison, though in B92 that makes no sense since Alice's basis is non-orthogonal. Instead, Bob publishes his classical results r for each of the measured qubits. He can do that over a public channel. All the values of r simply form a series of 0 and 1, each one corresponding to a specific qubit.

As we know, Alice could have encoded the data into the qubit in two different ways, as outlined in Eq. 6.4. For both of these options there are several possible measurement outcomes that might have lead Bob to obtain the r values that he actually got, so let us review them carefully:

- Alice generated $a = 0$ and thus encoded a state $|0\rangle$ into the qubit

  – Bob generates $b = 0$ and thus measures in the correct Z-basis

    · Bob measures $r = 0$. He cannot really say if Alice encoded $a = 0$, sent $|0\rangle$ and he measured in the correct basis, or whether she encoded $a = 1$, sent $|+\rangle$ and by measuring the qubit in the incorrect basis it simply collapsed to $|0\rangle$.
    · Bob cannot ever measure $r = 1$ when he guesses the basis right—in this case Alice always sends $|0\rangle$, and $r = 1$ would have meant here she sent $|1\rangle$

  – Bob generates $b = 1$ and thus measures in the incorrect X-basis

    · Bob measures $r = 0$. He cannot really say if Alice encoded $a = 1$, sent $|+\rangle$ and he measured in the correct basis, or whether she encoded $a = 0$, sent $|0\rangle$ and by measuring the qubit in the incorrect basis it simply collapsed to $|+\rangle$.
    · Bob measures $r = 1$. He is now certain that Alice generated $a = 0$, sent him $|0\rangle$ and it collapsed to $|-\rangle$ because he used the incorrect basis. The reason is, she would have never sent $|-\rangle$ in the X-basis.

- Alice generated $a = 1$ and thus encoded a state $|+\rangle$ into the qubit

  – Bob generates $b = 0$ and thus measures in the incorrect Z-basis

    · Bob measures $r = 0$. He cannot really say if Alice encoded $a = 0$, sent $|0\rangle$ and he measured in the correct basis, or whether she encoded $a = 1$, sent $|+\rangle$ and by measuring the qubit in the incorrect basis it simply collapsed to $|0\rangle$.
    · Bob measures $r = 1$. He is now certain that Alice generated $a = 1$, sent him $|+\rangle$ and it collapsed to $|1\rangle$ because he used the incorrect basis. The reason is, she would never sent $|1\rangle$ in the Z-basis.

  – Bob generates $b = 1$ and thus measures in the correct X-basis

    · Bob measures $r = 0$. He cannot really say if Alice encoded $a = 1$, sent $|+\rangle$ and he measured in the correct basis, or whether she encoded $a = 0$, sent $|0\rangle$ and by measuring the qubit in the incorrect basis it simply collapsed to $|+\rangle$.
    · Bob cannot ever measure $r = 1$ when he guesses the basis right—in this case Alice always sends $|+\rangle$, and $r = 1$ would have meant here she sent $|-\rangle$

Looking at all this combinations, we quickly recognize that the only possibility for Alice and Bob to be able to assert that they both have the same bits to make up their shared key, is in the situation when Bob's measurement result was $r = 1$. Therefore, they discard all the bits that correspond via their index to the qubits for which Bob measured $r = 0$.

The scheme also tells us that Bob's measurement result $r = 1$ could only occur when Bob guessed the basis wrong, so for qubits for which:

$$\{a = 0, b = 1, r = 1\} \tag{6.5}$$

$$\{a = 1, b = 0, r = 1\}$$

Those are the qubits that Alice and Bob need to keep to make up the secret key. The key for Alice will be the array of the remaining a bits, while for Bob it will be the array of negated b bits (or $1 - a$), because a and b are perfectly correlated but have opposite values, forcing Bob to flip the value of each of his b bits.

To sum up, this entire process results in the number of available bits going down from the initial $8n$ to about $2n$. If it is still not obvious, the simple way of thinking about it, is as follows. If Bob guesses the correct basis, the qubits will be discarded, which will cut their pool in half. From the remaining 50%, he will be measuring in the incorrect basis and so would obtain $r = 1$ only half of the time—when the qubit collapses to 1. This again means halving of the available qubit pool.

The overview of the B92 protocol, without the explicit eavesdropper check, is shown in Table 6.5.

With regards to the eavesdropper check, the procedure to follow in B92 would be identical to the one we discussed for BB84. After the reconciliation process we just covered, Alice and Bob still have about $2n$ bits remaining, which naturally means that they can deploy n of them as an integrity check against a presence of a potential malicious intermediary or other threat actors. What they need to engage in, is a comparison of n of their $\{a, b\}$ pairs, which can be executed over a public channel. Their expectation would be to find that the correlation between them—as explained earlier, they should have opposite values—remained intact, and if it does not, it would hint at a potential adversary on the quantum communication link. From the practical standpoint, this would be the trigger for Alice and Bob to abort the protocol and restart from scratch.

Overall, B92 provides an interesting alternative to BB84. By utilizing a single non-orthogonal basis, it benefits from a lower overall complexity level, at the price of its halved efficiency, which, at roughly 25% for BB84, was not too high to begin with. This leaves only 12.5% qubits to begin with error correction and privacy amplification [1], which will further impact the amount of usable transmitted data. These are topics we generally omit in this book, as we limit ourselves to idealistic theoretical conditions, but nevertheless they will be part of the bigger picture in real-life implementations.

It would not come as a surprise, when we say that the Q#-based sample B92 protocol implementation will largely follow the BB84 from Listing 6.2. We shall therefore often rely on and cross-reference code listings from that section.

Just like we did in BB84, we shall repeat the setup with a single operation that will be capable of orchestrating the entire protocol based on three input

Table 6.5 Overview of the B92 protocol

Alice bit a	$a = 0$			$a = 1$		
Alice state sent	$|0\rangle$			$|+\rangle$		
Bob bit b	$b = 0$	$b = 1$		$b = 0$		$b = 1$
Bob basis	Z	X		Z		X
Bob measurement r	0	0	1	0	1	0
Keep bit?	no	no	yes	no	yes	no

parameters—the total amount of roundtrips, the size of a roundtrip and the probability of an attacker sniffing out the communication between Alice and Bob. The operation RunB92Protocol (roundtrips : Int, rountripSize : Int, eavesdropperProbability : Double) in its entirety is shown in Listing 6.15.

```
operation RunB92Protocol(roundtrips : Int, rountripSize : Int,
    eavesdropperProbability : Double) : Unit {
    mutable aliceKey = [false, size = 0];
    mutable bobKey = [false, size = 0];
    for roundtrip in 0..roundtrips-1 {
        use qubits = Qubit[rountripSize];

        // 1. Alice chooses her random bits
        let aliceValues = GenerateRandomBitArray(rountripSize);

        // 2. Alice encodes the values in the qubits using the
            random bases
        EncodeQubits(aliceValues, qubits);

        // 3. Eve attempts to evesdrop based on the configurable
            probability
        Eavesdrop(eavesdropperProbability, qubits);

        // 4. Bob chooses his bases
        let bobBases = GenerateRandomBitArray(rountripSize);

        // 5. Bob measures qubits using the random bases
        let bobResults = DecodeQubits(bobBases, qubits);

        // 6. Alice and Bob perform the comparison and throw away
            unnecessary values
        let (aliceRoundTripResult, bobRoundTripResult)
            = ReconcileAliceAndBob(rountripSize, aliceValues,
                bobBases, bobResults);

        // 7. Append both key from this roundtrip to the overall
            key
        // repeat however many times needed
        set aliceKey += aliceRoundTripResult;
        set bobKey += bobRoundTripResult;
    }

    // 8. Perform the eavesdropper check
    let (errorRate, trimmedAliceKey, trimmedBobKey) =
        EavesdropperCheck(aliceKey, bobKey);

    // 9. Output the resulting keys and additional useful info
    ProcessResults(errorRate, trimmedAliceKey, trimmedBobKey,
        eavesdropperProbability);
}
```

Listing 6.15 Skeleton of the B92 implementation in Q#.

The code looks very similar to the BB84 orchestration operation from Listing 6.1, but there are some key differences. First of all, Alice does not need to keep track of her bases anymore, hence a dedicated array keeping a record of the random bases she chose is no longer needed—step one from BB84 is completely omitted. Instead, Alice proceeds straight to generating a random array of bits—an array of her *a* bits—which shall represent the bits she will send over to Bob to be used as the foundation for them establishing their shared key.

This procedure is performed by `GenerateRandomBitArray` (length : Int) operation, which is identical to the one from the BB84 protocol and which was already listed in Listing 6.2, so we do not need to revisit it again here. The randomly generated bits are stored in a local variable `aliceValues`.

Next step, step two in Listing 6.15, represents Alice encoding these randomly generated bits into the batch of qubits, one *a* bit into one qubit, that shall make up a given roundtrip. She does it with the help of an operation `EncodeQubits` (bits : Bool[], qubits : Qubit[]) from Listing 6.16, which is somewhat similar to the encoding operation from BB84, with the key difference being the usage of a single non-orthogonal basis, instead of two different independent bases.

If the bit value that Alice wants to encode is 0, she keeps the qubit as is, because as we already know, Q# guarantees that newly allocated qubits are in the computational basis state $|0\rangle$. Conversely, if she is encoding 1, she applies the **H** gate to create the state $|+\rangle$. The encoding operation mutates the qubits in place, and therefore no return from the operation is necessary.

```
operation EncodeQubits(bits : Bool[], qubits : Qubit[]) : Unit {
    for i in 0..Length(qubits)-1 {
        let valueSelected = bits[i];
        if (valueSelected) { H(qubits[i]); }
    }
}
```

Listing 6.16 Encoding of qubits step of the B92 protocol in Q#.

Step three in Listing 6.15 corresponds to step four in the BB84 in Listing 6.1, and it involves Eve mounting an attack on the communication channel between the qubits that are now (for our Q# code hypothetically, of course) in transit between Alice and Bob. Eve will be allowed to tap into the qubit communication between Alice and Bob based on the probability value which is an argument on our protocol operation. The implementation here is identical to the one from BB84 protocol as shown in Listing 6.4, so we will not repeat it. Since Eve does not know which measurement basis to use too, she will have to measure the qubits by guessing between the Z and X-bases.

Next part of the protocol requires Bob to receive the qubits, generate a random bit array corresponding to the measurement bases Z and X, and perform the measurements on the qubits accordingly. For our implementation of the B92, these are steps four and five in Listing 6.15, and they happen to be identical as steps five and six from the BB884 implementation in Listing 6.1. Namely, Bob shall generate his random bit array using the `GenerateRandomBitArray` (length : Int) operation as used in BB84 Listing 6.2, and then decode them the same way as in BB84, using `DecodeQubits` (bases : Bool[], qubits : Qubit[]) operation which we already saw in Listing 6.5.

Next, at step six of Listing 6.15, the process deviates from BB84 again. Alice and Bob should reconcile their data according to the procedure we outlined earlier—based on the measurements results that Bob obtained, which we deemed as r. If Bob measured $r = 1$, Alice should keep her random bit a, and Bob should keep his random base bit b. Since Bob knows that at this point a and b are perfectly correlated but inverses of each other, he also flips his bits to make sure that he produces an identical bit array to Alice. For all other cases, namely where $r = 0$, the corresponding a and b bits must be discarded. This is expressed in Q# code with the function shown in Listing 6.17.

```
function ReconcileAliceAndBob(length : Int, bitsAlice : Bool[],
    basesBob: Bool[], bitsBob : Bool[]) : (Bool[], Bool[]) {
    mutable aliceTrimmedBits = [];
    mutable bobTrimmedBits = [];

    for i in 0..length-1 {
        let r = bitsBob[i];
        let a = bitsAlice[i];
        let b = basesBob[i];

        if r {
            set aliceTrimmedBits += [a];
            set bobTrimmedBits += [not b];
        }
    }

    return (aliceTrimmedBits, bobTrimmedBits);
}
```

Listing 6.17 Alice and Bob reconciliation step of the B92 protocol in Q#.

The goal is to obtain two sets of trimmed bits representing a shared key in array data structures—aliceTrimmedBits for Alice and bobTrimmedBits for Bob. Those are then returned from the function as a tuple. Notice that because this callable does not perform any quantum work, it does not need to be an operation, but is declared as a function instead. The usage of the inline variables r, a and b is not necessary—we could have written the code without the explicit variable allocations, by relying on indexer accesses only, though this way it is a little more verbose and hopefully easier to follow.

The final parts of B92 would again converge with BB84. Due to the fact that our roundtrip sizes are relatively small, we will need to repeat the procedure outlined so far a larger amount of times so that the produced keys have sizes that would be usable for Alice and Bob.

The eavesdropper check operation, step eight in Listing 6.15, is the same as in BB84, and we covered it in Listing 6.7. Obviously, it results in halving of the key that remains in possession of Alice and Bob. This concludes the protocol—what follows is a call to the helper function from Listing 6.8, which we are re-using from earlier for providing the output of the B92 protocol execution.

The only thing left to do at this point, is to run the protocol using the entry operation from Listing 6.15. We want to do this with different configurations and verify that its outcome is as expected—that identical keys are shared between the parties and eavesdropper presence is successfully detected where needed.

Let us first invoke it without any eavesdropper presence, with 64 roundtrips and a roundtrip batch size of 16. Notice that, as already discussed, the total amount of involved qubits must be twice as large as in BB84 in order to obtain the same final key size.

```
@EntryPoint()
operation Start() : Unit {
    RunB92Protocol(64, 16, 0.0);
}
```

Listing 6.18 Execution of the B92 protocol in Q# without any configured eavesdropper.

The output of this operation should resemble that from Listing 6.19.

```
Alice's key: 46758148433604515044263567686035 1522796979
Bob's key:   46758148433604515044263567686035 1522796979
Error rate: 0%
Running the protocol with eavesdropping probability 0 was
    successful.
```

Listing 6.19 Sample output from running the code from Listing 6.18.

Just as we experienced in BB84, once the eavesdropper is brought into the picture, we should see the failure of the protocol. We can actually invoke the protocol twice in a row in one execution, by setting the eavesdropper probabilities to 50 and 100%.

```
@EntryPoint()
operation Start() : Unit {
    RunB92Protocol(64, 16, 0.5);
    RunB92Protocol(64, 16, 1.0);
}
```

Listing 6.20 Execution of the B92 protocol in Q# with preconfigured eavesdropper probabilities.

The observed result should prove that we were able to sniff out the attacker's presence in both cases. Depending on how likely it was for the eavesdropper to participate in the process, the detected error rate might be higher or smaller. In either case, Alice and Bob should have aborted the protocol.

```
Alice's key: 1621886061465128193461585039680001053027141 11422371
Bob's key:   1559786638873210653083674979518837013489268 11467171
Error rate: 15.625%
Eavesdropper detected!
Running the protocol with eavesdropping probability 0.5 was
    unsuccessful.


Alice's key: 7728681265505279559099071734316686598191063923
Bob's key:   1679692892896491706419107279380654957414073073
Error rate: 32.71889400921659%
Eavesdropper detected!
Running the protocol with eavesdropping probability 1 was
    unsuccessful.
```

Listing 6.21 Sample output from running the code from Listing 6.20.

Overall, despite some larger deviations on individual runs with regards to key sizes and error rates—that are caused by the fact that efficiency of B92 is lower than BB84, which is particularly visible in the small roundtrip sizes we deal with here—the results should quite nicely align with those of the BB84 implementation we did earlier.

6.4 EPR-Based Quantum Key Distribution

BB84 and B92 both require physical transmission of qubits from Alice to Bob. This creates a natural attack vector, as the qubits in transit can be intercepted by a potential adversary. On the other hand, entanglement seems to be perfectly suited for distributing secret information such as encryption keys.

We are already well aware that two actors could share sets of maximally entangled particle pair upfront, each one making up a Bell state (EPR pairs). At that point they can head off their separate direction and no longer need to have a quantum communication channel available between themselves. In fact, if the entangled pairs are provided by a third party (Charlie), Alice and Bob are never even required to be in contact in the first place.

Later on, provided no decoherence effects happened and the entanglement between their shared pairs remains stable,[4] Alice and Bob can be guaranteed that upon their respective measurements, they will receive perfectly correlated measurement results. This already takes us far down the path of quantum key distribution.

The first EPR-based quantum key distribution protocol was proposed by Artur Ekert in 1991, who suggested that the phenomenon of entanglement could facilitate QKD without any associated *element of reality* [9]. In line with the naming schemes we encountered so far, Ekert's protocol became known as E91. A set of different spin-off variants followed it, giving rise to a general class of quantum key distribution protocols, the so-called EPR-based quantum key distribution protocols.

The scheme devised by Ekert addressed two major challenges of the BB84 and B92 protocols. First of all, by relying on EPR correlations, the need for a quantum communication channel between Alice and Bob, as the medium to share the key, is no longer required, provided they both have a reliable alternative way of obtaining the entangled particle pairs. And of course if there is no communication, there is no channel to attack for an adversary.

Secondly, the secret key is never known by any of the parties upfront. In both BB84 and B92, the initial random sequence of bits is created by Alice, encoded into qubits, shipped off to Bob, and they proceed to establish the shared key from that. In EPR-based quantum key distribution schemes, the secret does not really exist as an objective element of reality on either side of the protocol until both participants perform their measurements. From that perspective we can refer to EPR-schemes not as shared secret *distribution* but shared secret *generation* methods [8]. This is

---

[4] A big *if* indeed.

important because if there is no key, there is nothing to steal—even if the EPR pairs do travel through some quantum optical channel. On top of that, the randomness of the secret is now underpinned by the correctness of quantum mechanism, instead of a randomness promise that needs to be fulfilled by one of the parties (typically Alice) engaged in the protocol.

Such characteristics of the EPR-based quantum key distribution protocols also mean that they can benefit from techniques such as privacy amplification [7] and other transmission enhancements, which in turn leads to overall better performance and resistance to noise [6].

The EPR protocols, despite the lack of qubit exchange between Alice and Bob during the execution protocol, are of course not free of attack possibilities. If the entangled qubit pairs are obtained by Alice and Bob from a third party source, an attacker, Eve, could have impersonated a trusted authority (Charlie) responsible for producing those entangled pairs in the first place, or injected (swapped) its own qubits into the communication channel between Alice and Bob and that EPR pair source. Even if Alice and Bob produce the entangled pairs when they are physically present together at some point in time, it is still possible for Eve to mount an attack at a later time. For example, once Alice and Bob are separated from each other, Eve could still gain temporary access to either Alice's or Bob's part of the EPR pair and attempted to measure them—after all, in this scheme, access to one side is potentially enough.

Let us now go through the steps of an EPR-based quantum key distribution. Alice and Bob begin with a set of n entangled qubit pairs in their possession. Each one is controlling one of the qubits making up a given pair, with each pair in an identical initial maximally entangled Bell state $|\Phi^+\rangle$.

$$|\Phi^+\rangle = \frac{1}{\sqrt{2}}\Big(|00\rangle + |11\rangle\Big) \tag{6.6}$$

Due to the nature of maximally entangled states, which we covered in Chap. 5, regardless of whether Alice or Bob decides to measure their part of the entangled pair first, it will force the other qubit making up the pair to produce perfectly correlated measurement result upon its observation—as long as the chosen basis is the same for both of the measurements. In the discussed case, since the entangled state is $|\Phi^+\rangle$, it would really be the exact same measurement outcome—they would both produce classical 0 or they would both produce classical 1. Thus, a set of such entangled EPR pairs will, upon measurement, equip both participants with identical classical bit sequences. For opposite correlated Bell states $|\Psi^+\rangle$ and $|\Psi^-\rangle$, the generated bit sequences would be complementary to each other, and one of the parties would be required to perform a bitflip corrective action at the end of the protocol.

When Alice and Bob are ready to have their shared key created, each of them performs a measurement of the qubit in their control. Just like in BB84, they will randomly toggle between two bases—the computational Z-basis $\{|0\rangle, |1\rangle\}$ and the X-basis $\{|+\rangle, |-\rangle\}$. If their randomly chosen basis happens to be the same for the given EPR pair, they will indeed receive the same classical bit. If their bases are different, they will only get the same result 50% of the time.

Table 6.6 Overview of the EPR QKD protocol. For a simplified picture it is assumed that Alice measures first

Alice basis	Z				X			
Alice measurement	$|0\rangle$		$|1\rangle$		$|+\rangle$		$|-\rangle$	
Bob basis	Z	X	Z	X	Z	X	Z	X
Probability of correlation	1	0.5	1	0.5	0.5	1	0.5	1

After the completion of all measurements, as it was done in BB84, Alice and Bob need to compare the measurement bases. This does not require any special security measures and therefore they can perform this comparison over a public channel. They will then proceed to discard all the qubit pairs for which their bases were different, as the produced bit result is not deterministic. Statistically speaking, their bases' choices will agree about half of the time, and therefore about 50% of their qubits will need to be discarded. The remaining half of the qubits now forms a random key that each of them has in their possession.

The summary of the protocol, with an assumption that it is Alice who performs the first measurement (though this is not a constraint of the scheme) is shown in Table 6.6.

The eavesdropper check follows the same principles as both BB84 and B92. Alice and Bob will have to sacrifice half of their remaining qubits, further bringing down the protocol efficiency to 25%, leaving them with only a quarter of the entangled qubit pairs contributing to the actual shared secret key. Since the process in which the key emerges relies on random basis guesses, Eve would have had to guess the basis as well if she attempted to perform a silent measurement of any of the involved qubits at any point, and she would therefore impact the correlations between the EPR pair. If the discrepancy rate is higher than the arbitrary acceptable error rate, Alice and Bob would determine that there is a high probability of an eavesdropper presence and they should abort the protocol and restart with a new set of entangled particles.

A complementary check for the eavesdropper activity would be for the parties to perform a Bell inequality violation test [2]. Inequalities should always be violated if the protocol was successful executed—otherwise the attacker must have disrupted the system by introducing those missing *elements of reality* into it.

Our Q# implementation for EPR-based quantum key distribution will largely follow the pattern we already established so far in this chapter when discussing BB84 and B92 protocols. In this particular case, our main orchestrator operation will be shown in Listing 6.22 where it is called RunEPRQKDProtocol (expected KeyLength : Int, eavesdropperProbability : Double).

The roundtrip configuration parameters, familiar from BB84 and B92 Q# implementations are replaced here with a single expectedKeyLength. As we already learnt, contrary to those protocols, the EPR-based quantum key distribution does not require the exchange of qubits between the two participating parties. As such, "sending" of qubits in chunks between Alice and Bob can be replaced with actions sequentially executed against single qubits by them, which altogether makes the simulation of the protocol considerably less complicated. Since we know that the efficiency of the protocol is 25%, we shall allocated a total of $4 \times$ expectedKeyLength qubit pairs.[5]

```
operation RunEPRQKDProtocol(expectedKeyLength : Int,
    eavesdropperProbability : Double) : Unit {
    // we require 4 * n EPR pairs to produce a key of length n (
        on average)
    let requiredQubits = 4 * expectedKeyLength;
    mutable aliceRes = [false, size = 0];
    mutable bobRes = [false, size = 0];

    // 1. Alice and Bob choose their bases
    let aliceBases = GenerateRandomBitArray(requiredQubits);
    let bobBases = GenerateRandomBitArray(requiredQubits);

    for i in 0..requiredQubits-1 {

        // 2. create an entangled EPR pair
        use (aliceQ, bobQ) = (Qubit(), Qubit());
        PrepareEntangledState([aliceQ], [bobQ]);

        // 3. Eve attempts to evesdrop based on the configurable
            probability
        Eavesdrop(eavesdropperProbability, [bobQ]);

        // 4 Alice and Bob measure in their random bases
        // Randomize who measures first
        if (DrawRandomBool(0.5)) {
            set aliceRes += [DoMeasure(aliceBases[i], aliceQ)];
            set bobRes += [DoMeasure(bobBases[i], bobQ)];
        } else {
            set bobRes += [DoMeasure(bobBases[i], bobQ)];
            set aliceRes += [DoMeasure(aliceBases[i], aliceQ)];
        }
    }

    // 5. Compare bases
    let (aliceKey, bobKey)
        = CompareBases(aliceBases, aliceRes, bobBases, bobRes);

    // 6. Perform the eavesdropper check
    let (errorRate, trimmedAliceKey, trimmedBobKey) =
        EavesdropperCheck(aliceKey, bobKey);

    // 7. Output the resulting keys and additional useful info
    ProcessResults(errorRate, trimmedAliceKey, trimmedBobKey,
        eavesdropperProbability);
}
```

Listing 6.22 Skeleton of the E91 implementation in Q#.

---

[5] The efficiency will be 25% on average, while on individual runs it may be considerably lower or higher, especially on smaller sample sizes, as the random bases chosen by Alice and Bob may deviate from each other from run to run. For simplicity, we shall ignore these statistical deviations and proceed in the code naively, without accounting for those effects.

Unsurprisingly, a lot of the code steps outlined in Listing 6.22 already look familiar to the corresponding BB884 and B92 protocol code from Listings 6.1 and 6.15. That is indeed the case, and we will focus specifically on the differences, while reusing as much of the code that was already written and discussed as possible.

Similarly to BB84, Alice and Bob will need to operate on two bases in this QKD variant, and therefore classic bit arrays corresponding to those bases are generated upfront by calls to the familiar helper operation, GenerateRandomBitArray (length : Int), which we first introduced in Listing 6.2. A randomly drawn 0 will indicate the computational basis $\{|0\rangle, |1\rangle\}$, while a 1 will be mapped to the X-basis $\{|+\rangle, |-\rangle\}$. This happens in step one in Listing 6.22.

Next, as step two, an entangled pair in state $|\Phi^+\rangle$ is created using the built-in Q# standard library operation PrepareEntangledState (left : Qubit[], right : Qubit[]) operation, from the Microsoft. Quantum.Preparation namespace.

In step three of Listing 6.22, we execute the eavesdropper simulation, based on the configurable probability. The implementation of that is identical to the one used earlier, and the code comes from Listing 6.4. Note that in both BB84 and B92, we only dealt with one set of qubits, that moved from Alice to Bob. In the case of EPR-based quantum key distributions, the total amount of involved qubits is actually double—because both Alice *and* Bob have their own sets. This also means that the eavesdropper could potentially attempt to attack either of these two qubit collections. Our implementation in Listing 6.22 arbitrarily decides that the eavesdropper is spying on Bob's qubit, by explicitly choosing his bobQubit as input into the eavesdropping simulation operation, but in reality the same results would occur if we performed the spying activities on either just Alice's qubits, or, in fact, on both Alice's and Bob's qubits together.

The next thing for Alice and Bob to do, is to measure their qubits in their random bases, which is marked as step four in Listing 6.22. The order of measurement does not matter for EPR correlations, but to be able to verify that, we perform measurement order randomization on each qubit to make that very explicit. In real life, Alice and Bob would be measuring while being spatially separated, most likely without any particularly strict time coordination.

Alice and Bob measure their qubits using a helper DoMeasure (basisBit : Bool, qubit : Qubit), which takes care of switching between the bases based on the random basis bits. It returns a classical bit result, which is the appropriately allocated to either aliceRes or bobRes array. The operation is shown in Listing 6.23.

```
operation DoMeasure(basisBit : Bool, qubit : Qubit) : Bool {
    // - 0 will represent {|0>,|1>} computational (PauliZ) basis
    // - 1 will represent {|->,|+>} Hadamard (PauliX) basis

    let result = Measure([basisBit ? PauliX | PauliZ], [qubit]);
    let classicalResult = ResultAsBool(result);
    return classicalResult;
}
```

Listing 6.23 Measurement step of the E91 protocol implementation in Q#.

The remaining steps five to seven in Listing 6.22 are identical to their counterparts from the BB84 protocol in Listing 6.1.

First, Alice and Bob engage in comparing their bases and throwing away those results that came from qubits that were measured in non-matching bases. This is done with the CompareBases (basesAlice : Bool[], bitsAlice : Bool[], basesBob : Bool[], bitsBob : Bool[]) operation, which is the same as the one from Listing 6.6. This trims the length of the results arrays for both Alice and Bob by roughly 50%. Next, they perform the eavesdropper check, which is the same as used in our BB84 and B92 implementations, and uses the operation defined in Listing 6.7, which again halves the size of the resulting bit sequences. Finally, the protocol execution details and summary are printed out for our convenience using the post processing helper function which was first declared in Listing 6.8.

Testing of our implementation happens by invoking the main entry point operation from Listing 6.22. As in other sections in this chapter, we shall verify it using parameterized eavesdropper presence set to 0, 50 and 100%. The test key length will be 128 bits, meaning we will need to use a total of 512 qubit pairs.

```
@EntryPoint()
operation Start() : Unit {
    RunEPRQKDProtocol(128, 0.0);
    RunEPRQKDProtocol(128, 0.5);
    RunEPRQKDProtocol(128, 1.0);
}
```

Listing 6.24 Execution of the E91 protocol in Q# with various preconfigured eavesdropper probabilities.

The output of the program will cover all three eavesdropper variants and should be roughly similar to the one shown in Listing 6.25—with, as expected, only the execution without eavesdropper being considered successful.

```
Alice's key as int: 82518729576956262613543945806793059258910070001
Bob's key as int:   82518729576956262613543945806793059258910070001
Error rate: 0%
Running the protocol with eavesdropping probability 0 was
    successful.

Alice's key as int: 61101240891429193942196596614735175587949
Bob's key as int:   148298585159193065860217011824224380622949
Error rate: 13.91304347826087%
Eavesdropper detected!
Running the protocol with eavesdropping probability 0.5 was
    unsuccessful.

Alice's key as int: 48836220763471090461317447127383126496430
Bob's key as int:   35016402630204727728442566339165573337386
Error rate: 24.793388429752067%
Eavesdropper detected!
Running the protocol with eavesdropping probability 1 was
    unsuccessful.
```

Listing 6.25 Sample output from running the code from Listing 6.24.

References

1. Abu-Ayyash, A. M., & Ajlouni, N. (2008). QKD: Recovering unused quantum bits. *Pages 1–5 of: 2008 3rd International Conference on Information and Communication Technologies: From Theory to Applications.*
2. Acin, A., Gisin, N., & Masanes, L. (2006). From Bell's theorem to secure quantum key distribution. *Physical Review Letters, 97*(12).
3. Bennett, C. H. (1992). Quantum cryptography using any two nonorthogonal states. *Physical Review Letters, 68*(May), 3121–3124.
4. Bennett, C. H., & Brassard, G. (2014). Quantum cryptography: Public key distribution and coin tossing. *Theoretical Computer Science, 560*(Dec), 7–11.
5. Boaron, A., Boso, G., Rusca, D., Vulliez, C., Autebert, C., Caloz, M., Perrenoud, M., Gras, G., Bussières, F., Li, M.-J., Nolan, D., Martin, A., & Zbinden, H. (2018). Secure quantum key distribution over 421 km of optical fiber. *Physical Review Letters, 121*(Nov), 190502.
6. Bruß, D., & Lütkenhaus, N. (1999). Quantum key distribution: From principles to practicalities. *Applicable Algebra in Engineering, Communications and Computing, 10*(02).
7. Deutsch, D., Ekert, A., Jozsa, R., Macchiavello, C., Popescu, S., & Sanpera, A. (1996). Quantum privacy amplification and the security of quantum cryptography over noisy channels. *Physical Review Letters, 77*(Sep), 2818–2821.
8. Djordjevic, I. B. (2019). *Physical-layer security and quantum key distribution.* Springer Nature Switzerland.
9. Ekert, A. K. (1991). Quantum cryptography based on Bell's theorem. *Physical Review Letters, 67*(Aug), 661–663.
10. Gibney, E. (2016). Chinese satellite is one giant step for the quantum internet. *Nature, 535*(7613), 478–479.
11. Hughes, R. J., Morgan, G. L., & Peterson, C. G. (2000). Quantum key distribution over a 48 km optical fibre network. *Journal of Modern Optics, 47*(2–3), 533–547.
12. Liao, S.-K., Cai, W.-Q., Handsteiner, J., Liu, B., Yin, J., Zhang, L., Rauch, D., Fink, M., Ren, J., Liu, W.-Y., et al. (2018). Satellite-relayed intercontinental quantum network. *Physical Review Letters, 120*(3).
13. Lo, H.-K., & Zhao, Y. (2008). *Quantum Cryptography.*
14. Nielsen, M. A., & Chuang, I. (2010). *Quantum computation and quantum information: 10th anniversary edition.* Cambridge University Press.
15. Scarani, V., Acín, A., Ribordy, G., & Gisin, N. (2004). Quantum cryptography protocols robust against photon number splitting attacks for weak laser pulse implementations. *Physical Review Letters, 92*(5).
16. Stucki, D., Legré, M., Buntschu, F., Clausen, B., Felber, N., Gisin, N., Henzen, L., Junod, P., Litzistorf, G., Monbaron, P., et al. (2011). Long-term performance of the SwissQuantum quantum key distribution network in a field environment. *New Journal of Physics, 13*(12), 123001.
17. Vcking, B., Alt, H., Dietzfelbinger, M., Reischuk, R., Scheideler, C., Vollmer, H., & Wagner, D. (2011). *Algorithms unplugged* (1st ed.). Springer Publishing Company, Incorporated.
18. Zhao, S.-C., Han, X.-H., Xiao, Y., Shen, Y., Gu, Y.-J., & Li, W.-D. (2019). Performance of underwater quantum key distribution with polarization encoding. *The Journal of the Optical Society of America A, 36*(5), 883–892.

Part III

Chapter 7
Algorithms

In the final chapter of this book we are going to take all of the knowledge accumulated so far and put it all to good use—as we walk through some of the most important quantum algorithms.

This chapter will also, inevitably, be a celebration of the amazing theoretical achievements of some of the greatest minds in the field of quantum computing, and a chance for us to really appreciate their incredible insight, as well as mathematical beauty and innovative thinking behind their work.

7.1   Deutsch–Jozsa Algorithm

In computability theory a black-box function is one that has unknown implementation, and can only be reasoned about using its inputs and outputs. A related concept, commonly used in computer science and in complexity theory, is one of an oracle, which can be constructed to study the black-box function and its decision tree.

7.1.1   Deutsch's Problem

In 1985, David Deutsch formulated his landmark black-box problem [9]. He imagined an unknown function, which takes a single bit input and produces a single bit output:

$$f : \{0, 1\} \to \{0, 1\} \tag{7.1}$$

Such function can be fully described by four possible cases, as outlined in Table 7.1. Deutsch then asked how many queries are needed to determine if the

© The Author(s), under exclusive license to Springer Nature Switzerland AG 2022   215
F. Wojcieszyn, *Introduction to Quantum Computing with Q# and QDK*, Quantum Science and Technology, https://doi.org/10.1007/978-3-030-99379-5_7

Table 7.1 Possible configurations for the Deutsch's problem black-box function

	x = 0	x = 1	Type
f(x)	0	0	Constant
f(x)	1	1	Constant
f(x)	1	0	Balanced
f(x)	0	1	Balanced

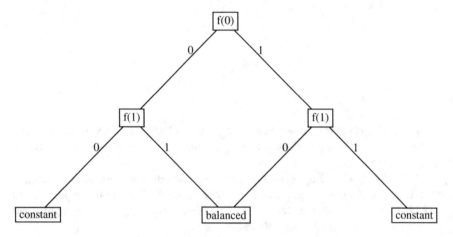

Fig. 7.1 Deutsch algorithm decision tree

function is *constant* or *balanced*, where we would consider a function to be constant if $f(0) = f(1)$, while it would be balanced if $f(0) \neq f(1)$.

At first glance the problem stated by Deutsch does not appear particularly interesting or, as matter of fact, challenging. However, it constituted a massive leap forward in both computation and complexity theories, as it marked the first known problem definition that could objectively be solved in a more efficient way on quantum computational device, compared to a classical computer.

Given an unknown implementation of the function, the classical decision tree for Deutsch's problem can be broken down into two separate steps, as shown in Fig. 7.1. It is irrelevant whether we first invoke $f(0)$ and then $f(1)$ or vice versa—both case evaluations are needed to determine whether the function is constant or balanced.

In other words, two logical queries to the oracle are required to solve the answer. Naturally, this reasoning aligns well with our intuition, which in turn simply reflects what one could call common sense. The spectacular insight of David Deutsch was that by taking advantage of a combination of superposition and interference effects of quantum mechanics, a quantum computer could provide the answer to the same problem in just a single step.

We already know that computations on a quantum computer must obey the laws of quantum mechanics, and one of the primary rules is that any quantum state transformation must be unitary, and thus reversible. On the other hand, the black-box function $f(x)$ is not possible to be directly modelled using a single qubit on quantum hardware, becuase in the cases where it is constant it is not reversible. A clever solution to this is to not only construct the function with a single qubit corresponding to the single bit on which the function operates, but to use an extra qubit to facilitate reversibility.

Suppose we call our oracle $\mathbf{U_f}$, then a general description of an oracle function would be

$$\mathbf{U_f}\,|x\rangle = |x \oplus f(x)\rangle \tag{7.2}$$

Taking into account the reversibility requirement, this needs be extended to two qubits—as a result it requires at least a two-qubit quantum computer to run.

$$\mathbf{U_f}(|x\rangle\,|y\rangle) = |x\rangle\,|y \oplus f(x)\rangle \tag{7.3}$$

To prove that reversibility is satisfied in such transformation scheme, we can take the output of $\mathbf{U_f}$ and apply $\mathbf{U_f}$ to it. This is shown in Eq. (7.4).

$$\mathbf{U_f}(|x\rangle\,|y \oplus f(x)\rangle) = \tag{7.4}$$
$$|x\rangle\,|(y \oplus f(x)) \oplus f(x)\rangle = |x\rangle\,|y \oplus (f(x) \oplus f(x))\rangle = |x\rangle\,|y \oplus 0\rangle =$$
$$|x\rangle\,|y\rangle$$

The quantum circuit to solve Deutsch's problem in a single logical query, which can be used to construct a corresponding quantum program, is shown in Fig. 7.2. It is not identical to the solution originally suggested by Deutsch [9] as that one was probabilistic—it would provide the expected answer, "constant" or "balanced", only half of time, while the other half it would produce a result "inconclusive". The implementation variant we describe here is an improvement to Deutsch's solution which would return deterministic result in a single query, and which was published by Cleve et al. [6].

The actual input value into the oracle is carried by the first qubit (the x), while the second qubit will carry the reversible result ($|y \oplus f(x)\rangle$). The extra qubit required by some of the quantum algorithms in order to satisfy reversibility requirements is referred to as *auxiliary* qubit.

We shall deal with the internals of the $\mathbf{U_f}$ oracle in a moment, but for now let us focus on the rest of the circuit. We shall call the initial state of the system $|\psi_1\rangle$ and

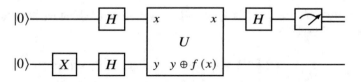

Fig. 7.2 Deutsch circuit

it consists of two qubits in the default $|0\rangle$ state.

$$|\psi_1\rangle = |00\rangle \tag{7.5}$$

The next step is apply the bitflip **X** transformation to the second qubit.

$$|\psi_2\rangle = (\mathbf{I} \otimes \mathbf{X}) |00\rangle = |01\rangle \tag{7.6}$$

Then an Hadamard transformation is performed on both of the qubits making up the circuit. The resulting state $|\psi_2\rangle$ will be the input to the $\mathbf{U_f}$ oracle.

$$|\psi_3\rangle = (\mathbf{H} \otimes \mathbf{H}) |01\rangle = \frac{1}{\sqrt{2}} \left(|0\rangle + |1\rangle \right) \frac{1}{\sqrt{2}} \left(|0\rangle - |1\rangle \right) \tag{7.7}$$

We can quickly realize that these are of course nothing else than states $|+\rangle$ and $|-\rangle$ respectively.

$$|\psi_3\rangle = |+\rangle |-\rangle \tag{7.8}$$

Since we already mentioned that the first qubit, currently in state $|+\rangle$, is responsible for the input into the oracle, let us step back for a moment, forget that we already have it in state $|+\rangle$ and treat it as a generic state $|x\rangle$, which would allow us to easier reason about the underlying concept being relied on here. Applying the oracle $\mathbf{U_f}$ to the such generalized state $|x\rangle |-\rangle$ produces the following

$$\mathbf{U_f} |x\rangle |-\rangle = \mathbf{U_f} |x\rangle \frac{1}{\sqrt{2}} \left(|0\rangle - |1\rangle \right) = |x\rangle \frac{1}{\sqrt{2}} \left(|0 \oplus f(x)\rangle - |1 \oplus f(x)\rangle \right) =$$
$$(-1)^{f(x)} |x\rangle |-\rangle \tag{7.9}$$

As it turns out, this algebraic trickery is commonly used in quantum information theory. Not only did it simplify our calculations, but it allowed us to encode the information about the function evaluation, $f(x)$, into the global phase. This technique is therefore called a *phase kickback*.

We can now substitute $|x\rangle$ with $|+\rangle$. Thanks to that, after evaluating the oracle against input state $|+\rangle |-\rangle$, we can describe the state of our system, now at $|\psi_4\rangle$, as:

$$|\psi_4\rangle = \mathbf{U_f} |+\rangle |-\rangle = \frac{1}{2} \left((-1)^{f(0)} |0\rangle + (-1)^{f(1)} |1\rangle \right) (|0\rangle - |1\rangle) \tag{7.10}$$

Obviously $(-1)^0 = 1$ and $(-1)^1 = -1$, so we can deduce that we have two outcome possibilities here:

$$|\psi_4\rangle = \begin{cases} \pm \frac{1}{2} (|0\rangle + |1\rangle)(|0\rangle - |1\rangle) & \text{when f(0) = f(1)} \\ \pm \frac{1}{2} (|0\rangle - |1\rangle)(|0\rangle - |1\rangle) & \text{when f(0)} \neq \text{f(1)} \end{cases} \tag{7.11}$$

This is already very close to obtaining the result—the final step is to apply the
$\mathbf{H} \otimes \mathbf{I}$ transformation to $|\psi_4\rangle$. We know that $\mathbf{H}$ is its own inverse so it will transform
the first qubit back into the Z-basis eigenstate $|0\rangle$ or $|1\rangle$.

$$|\psi_5\rangle = (\mathbf{H} \otimes \mathbf{I}) |\psi_4\rangle = \begin{cases} \pm |0\rangle \, (|-\rangle) \text{ when } f(0) = f(1) \\ \pm |1\rangle \, (|-\rangle) \text{ when } f(0) \neq f(1) \end{cases} \tag{7.12}$$

The state of the first qubit, when measured, will reveal the answer to the problem
for us. The second qubit stays in state $|-\rangle$ and is uninteresting to us—as defined in the
circuit Fig. 7.2, we do not even need to measure it. The conclusion is therefore that:

$$|\psi_5\rangle = |f(0) \oplus f(1)\rangle \, (|-\rangle) \tag{7.13}$$

This is a spectacular achievement, because we managed to determine $f(0) \oplus f(1)$
in a single function evaluation. This is of course something that a classical computer
would always need two logical queries for, and which is an excellent example of the
concept of *quantum parallelism*. As defined in Eq. (7.12), if the measurement of the
main qubit, which at this point is going to have a deterministic result, produces $|0\rangle$,
we deal with a constant function, and when it produces $|1\rangle$, we have a balanced one. It
is however worth emphasizing that to solve Deutsch's problem a quantum computer
does not actually perform different function evaluations in parallel, but instead relies
on the interference phenomenon to let incorrect answers cancel themselves out.

We can also notice one final thing with the procedure of solving the Deutsch's
problem—the measurement follows the $\mathbf{H}$ transformation, therefore it can also con-
ceptually be thought of as measurement of the output of the oracle—state $|\psi_4\rangle$—in
the X, instead of Z-basis.

7.1.2 Deutsch–Jozsa Generalization

In 1992 David Deutsch published a follow-up paper [10], co-authored with Richard
Jozsa, which provided a generalization of the original simple Deutsch's algorithm.
Thus, before we dive into the usual Q# implementation of the theory that was just
covered, let us first shift our attention to this generalized variant.

The Deutsch–Jozsa algorithm extends the original problem from a function with
a single bit input, to one that can accept n bits as arguments, such that:

$$f : \{0, 1\}^n \to \{0, 1\} \tag{7.14}$$

The problem is then still formulated as having to determine if the function is constant, or whether it is balanced, with the condition that balanced means that for exactly
half of the possible x input arguments the function returns 0, while returning 1 for the
other half. The consequence of such formulation is that according to classical computation theory, in the worst case scenario, as many as $2^{n-1} + 1$ function evaluations

may be need to obtain the provide the answer to the problem with 100% certainty. On the other hand, using the principles of interference and quantum parallelism, a quantum computer can still solve this in a single logical oracle evaluation.

In terms of defining the Deutsch–Jozsa problem as a black-box unitary transformation $\mathbf{U_f}$, we would say it still performs a transformation

$$\mathbf{U_f}(|x\rangle\,|y\rangle) = |x\rangle\,|y \oplus f(x)\rangle \tag{7.15}$$

with the qualification that now $x \in \{0, 1\}^n$ and $f(x) \in \{0, 1\}$. In other words, $|x\rangle$ no longer corresponds to a single qubit, but instead can span n qubits. The entire algorithm would then follow the exact same reasoning as the single bit variant, with the only difference being an extended input qubit register in play, and thus the relevant operations, such as the $\mathbf{H}$ transformations, are applied to this entire n qubit register. The quantum circuit for the algorithm, a slightly modified version of Fig. 7.2, is shown in Fig. 7.3.

The circuit starts with a qubit register of n qubits, and the auxiliary qubit, all in the default state $|0\rangle$.

$$|\psi_1\rangle = |0\rangle^{\otimes n}\,|0\rangle \tag{7.16}$$

This is then followed by flipping the auxiliary qubit, like we did earlier, using the $\mathbf{X}$ transformation.

$$|\psi_2\rangle = (\mathbf{I}^{\otimes n} \otimes \mathbf{X})\,|\psi_1\rangle = |0\rangle^{\otimes n}\,|1\rangle \tag{7.17}$$

Next, both the n qubit register and the auxiliary qubit are placed in a uniform superposition using the $\mathbf{H}$ transformation. Applying $\mathbf{H}^{\otimes n}$ to an n size register is of course nothing else but the Walsh–Hadamard transformation that we already covered in Sect. 4.3. As a reminder

$$\mathbf{H}^{\otimes n}\,|00\ldots 0\rangle = \frac{1}{\sqrt{2^n}} \sum_{x=0}^{2^n-1} |x\rangle \tag{7.18}$$

Therefore the next state $|\psi_3\rangle$ is

$$|\psi_3\rangle = (\mathbf{H}^{\otimes n} \otimes \mathbf{H})\,|0\rangle^{\otimes n}\,|1\rangle = \frac{1}{\sqrt{2^n}} \sum_{x=0}^{2^n-1} |x\rangle\, \frac{1}{\sqrt{2}}\Big(|0\rangle - |1\rangle\Big) \tag{7.19}$$

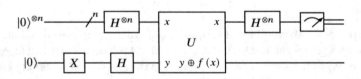

Fig. 7.3 Deutsch–Jozsa circuit

For brevity we can rewrite the state of the auxiliary qubit into $|-\rangle$.

$$|\psi_3\rangle = \frac{1}{\sqrt{2^n}} \sum_{x=0}^{2^n-1} |x\rangle |-\rangle \tag{7.20}$$

Upon invoking the black-box oracle $\mathbf{U_f}$, the state changes to:

$$|\psi_4\rangle = \mathbf{U_f} |\psi_3\rangle = \frac{1}{\sqrt{2^n}} \sum_{x=0}^{2^n-1} (-1)^{f(x)} |x\rangle |-\rangle \tag{7.21}$$

Finally, we apply the Walsh–Hadamard $\mathbf{H}^{\otimes n}$ transformation to the $|x\rangle$ qubit registry again. From this point on we will also no longer need to be interested in the auxiliary qubit—it remains, as before, in state $|-\rangle$:

$$|\psi_5\rangle = (\mathbf{H}^{\otimes n} \otimes \mathbf{I}) |\psi_4\rangle = \frac{1}{\sqrt{2^n}} \sum_{x=0}^{2^n-1} (-1)^{f(x)} \cdot \frac{1}{\sqrt{2^n}} \sum_{y=0}^{2^n-1} (-1)^{x \cdot y} |y\rangle |-\rangle \tag{7.22}$$

where

$$x \cdot y = x_0 y_0 \oplus x_1 y_1 \oplus \cdots \oplus x_{n-1} y_{n-1} \tag{7.23}$$

A simplified variant of $|\psi_5\rangle$ can be achieved by rearranging the terms a little, and dropping the auxiliary qubit.

$$\frac{1}{2^n} \sum_{y=0}^{2^n-1} \left(\sum_{x=0}^{2^n-1} (-1)^{x \cdot y + f(x)} \right) |y\rangle \tag{7.24}$$

This is now a state that is ready for measurement, but what is its significance? The probability amplitude for measuring $|0\rangle^{\otimes n}$ is

$$\frac{1}{2^n} \sum_{x=0}^{2^n-1} (-1)^{f(x)} \tag{7.25}$$

because for such case $|y\rangle = |0\rangle^{\otimes n}$, so $y = 0$.

If the function $f(x)$ was constant, the above probability amplitude reduces to 1 (when $f(x) = 0$) or to -1 (when $f(x) = 1$). In either case, the classical probability—a squared amplitude—for reading out $|0\rangle^{\otimes n}$ is equal to one when we deal with a constant function. To phrase it differently, when a joint measurement of the main register returns 0 for all of the qubits, we know with certainty that the black-box operation was constant, in just a single operation.

Conversely, the probability amplitude (7.25) will reduce to 0 when the function is balanced because the terms cancel each other out—we have equally many positive and negative terms in the sum. As a result, for a balanced function, the classical probability of a joint measurement result to be $|0\rangle^{\otimes n}$ becomes also zero. The consequence is of course that as soon as even a single of the measured qubits in the register produces a measurement result of 1, it reveals to us that the function $f(x)$ was balanced.

Despite dealing with quite an artificial and not particularly useful problem, the Deutsch–Jozsa algorithm is quite a stunning example of a possible computational speed-up that quantum devices may deliver. In this variant of the problem, a quantum solution still provides an answer in a single logical query, compared to possibly as many as $2^{n-1} + 1$ function evaluations using a classical computational device.

7.1.3   Oracle Definitions

The theoretical discourse thus far has led us to the conclusion that both Deutsch and Deutsch–Jozsa variants of the problem can indeed be efficiently solved on a quantum hardware. That said, in order to fully implement the solution suitable for execution on a quantum device, we still need to model the four different variants of the black-box $\mathbf{U_f}$ function. We shall refer to them as f_{c1} and f_{c2}, for the constant functions, and f_{b1} and f_{b2} for the balanced ones.

For the basic Deutsch problem, given that we only operate on a single bit, we can easily list all possible permutations (in fact we already did in Table 7.1). The functions behave as follows:

- $f_{c1}(0) = f_{c1}(1) = 0$
- $f_{c2}(0) = f_{c2}(1) = 1$
- $f_{b1}(0) = 0 \quad f_{b1}(1) = 1$
- $f_{b2}(0) = 1 \quad f_{b2}(1) = 0$

For the more sophisticated Deutsch–Jozsa formulation, the input is now x of n bit length. The constant functions are still the same as before, but for the balanced ones the situation becomes a little more complicated. For convenience of implementation we can approach this by treating the balanced variants as being driven by an XOR between all of the input bits. This would then mean that our balanced $f(x)$ returns 0 unless x an odd number of 1s is found in x, in which case it returns 1.

- $f_{c1}(x_0 \ldots x_{(n-1)}) = 0$
- $f_{c2}(x_0 \ldots x_{(n-1)}) = 1$
- $f_{b1}(x_0 \ldots x_{(n-1)}) = 0$ when $x_0 \oplus x_0 \oplus \cdots \oplus x_{(n-1)} = 0$
- $f_{b1}(x_0 \ldots x_{(n-1)}) = 1$ when $x_0 \oplus x_0 \oplus \cdots \oplus x_{(n-1)} = 1$
- $f_{b2}(x_0 \ldots x_{(n-1)}) = 1$ when $x_0 \oplus x_0 \oplus \cdots \oplus x_{(n-1)} = 0$
- $f_{b2}(x_0 \ldots x_{(n-1)}) = 0$ when $x_0 \oplus x_0 \oplus \cdots \oplus x_{(n-1)} = 1$

We already know that Deutsch formulation is simply a special case of Deutsch–Jozsa generalization, namely one where $n = 1$. Therefore it is enough for us to model

Fig. 7.4 Deutsch–Jozsa algorithm $\mathbf{U_f}$ implementation for f_{c1}

Fig. 7.5 Deutsch–Jozsa algorithm $\mathbf{U_f}$ implementation for f_{c2}

Fig. 7.6 Deutsch–Jozsa algorithm $\mathbf{U_f}$ implementation for f_{b1}

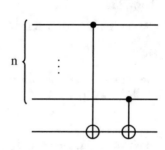

only the Deutsch–Jozsa specific functions using quantum circuits, as it would cover all the cases anyway.

In particular, the constant functions would be very simple. f_{c1} always produces 0, hence we only need to apply $\mathbf{I}^{\otimes n} \otimes \mathbf{I}$ transformation to the qubits which are passed into the oracle, as shown in Fig. 7.4. This, naturally, is semantically equivalent to doing nothing. Notice that we specifically do not mark on the circuit which state the input qubits will be in when $\mathbf{U_f}$ executes—instead we are only concerned with the implementation of its function body.

Conversely, f_{c2} would flip the result bit, which on a quantum circuit would mean executing the $\mathbf{X}$ transformation on the auxiliary qubit—$\mathbf{I}^{\otimes n} \otimes \mathbf{X}$. This is visualized in Fig. 7.5.

Since we already established that balanced functions would rely on XOR logic, they can be implemented using a series of **CNOT** gates. f_{b1} would is pictured in Fig. 7.6.

It's twin, f_{b2}, would be identical, except for an extra $\mathbf{X}$ applied to the auxiliary qubit. This is shown in Fig. 7.7.

7.1.4 Q# Implementation

Equipped with all of this, we can now proceed to the Q# implementation of the algorithms. Since we were just defining the $\mathbf{U_f}$ architecture using quantum circuits,

Fig. 7.7 Deutsch–Jozsa
algorithm U_f
implementation for f_{b2}

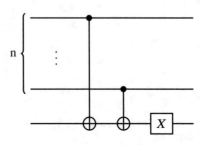

we might as well start the Q# code with them. The circuits are not particularly complicated so the corresponding code is also relatively simple. All four of them are shown in Listing 7.1.

```
operation ConstantZero(x : Qubit[], y : Qubit)
 : Unit is Adj {
}

operation ConstantOne(x : Qubit[], y : Qubit)
 : Unit is Adj   {
    X(y);
}

operation BalancedEqual(x : Qubit[], y : Qubit)
 : Unit is Adj   {
    for qubit in x {
        CNOT(qubit, y);
    }
}

operation BalancedNotEqual(x : Qubit[], y : Qubit)
 : Unit is Adj {
    for qubit in x {
        CNOT(qubit, y);
    }
    X(y);
}
```

Listing 7.1 All four Deutsch–Jozsa algorithm U_f implementations in Q#.

Notice that for the cases where we applied **I** in the circuits, we can simply leave the Q# operations doing nothing. An explicit mention of **I** helps on the circuit and in algebraic model, but at runtime has not impact. The return type of the operations can be Unit. All of the qubits that act as input are managed by the code outside of these operations anyway, so they are all transformed in place and effectively act as function output.

Next, comes the core part of the algorithm, which will follow the procedure outlined in Fig. 7.3. The operation implementing the algorithm is defined in a way where we can pass in two parameters—the n size and a reference to any of the four U_f variants, represented by a delegate, and is shown in Listing 7.2.

```
operation DeutschJozsaAlgorithm(n : Int, oracle :
    ((Qubit[], Qubit) => Unit)) : Bool {
    use (x, y) = (Qubit[n], Qubit());
    X(y);
    ApplyToEach(H, x);
    H(y);

    oracle(x, y);

    ApplyToEach(H, x);

    let allZeroes = All(ResultIsZero, MultiM(x));
    Reset(y);
    return allZeroes;
}

function ResultIsZero(result : Result) : Bool {
    return result == Zero;
}
```

Listing 7.2 Deutsch–Jozsa algorithm implementation in Q#.

Internally, the code would allocate the necessary qubits—both the core n-size qubit register and the auxiliary qubit. Though they do not necessarily have to be split into separate variables, for readability purposes, the register is represented by a qubit array x, of size n, while the auxiliary qubit is separate from it under the y variable. The subsequent steps follow the flow of the algorithm that we already covered, until we reach the measurement process. At that point we need to find out whether all qubits measure to $|0\rangle$—if yes, we deal with a constant function, otherwise the function would be a balanced one. The measurement is based on the MultiM (targets : Qubit[]) operation, which conveniently measures an array of qubits into an array of Result, in the computational basis. Because the auxiliary qubit y is never measured, it needs to be manually reset.

Overall such structure of the operation will allow us to invoke this algorithm for all four black-box functions, and thanks to the parameterized n, it will also support testing both the basic Deutsch algorithm for a single bit, and the Deutsch–Jozsa generalization for larger input bit sizes.

The final thing needed is the orchestration code to execute all of these cases, which is shown in Listing 7.3. It covers eight possible cases, four for Deutsch problem, and four for Deutsch–Jozsa variant, with the latter invoked for $n = 6$ bits. The usage of a loop in this orchestrator facilitates easy testing of the code against other values of n as well. The algorithm result is ultimately converted into string form, which will be equal to either "constant" or "balanced", and which gets printed to the console output.

```
@EntryPoint()
operation Main() : Unit {
    for n in [1, 6] {
        let constZero = DeutschJozsaAlgorithm(n, ConstantZero);
        let constOne = DeutschJozsaAlgorithm(n, ConstantOne);
        let balEq = DeutschJozsaAlgorithm(n, BalancedEqual);
```

```
    let balNotEq = DeutschJozsaAlgorithm(n, BalancedNotEqual);

    Message($"n={n}");
    Message($"Constant 0. Result: {Format(constZero)}.");
    Message($"Constant 1. Result: {Format(constOne)}.");
    Message($"Balanced. Result: {Format(balEq)}.");
    Message($"Balanced opp. Result: {Format(balNotEq)}.");
    }
}

function Format(result : Bool) : String {
    return result ? "constant" | "balanced";
}
```
Listing 7.3 Deutsch–Jozsa algorithm Q# orchestration code.

After executing this program, we shall see the following output in the console, confirming our expectations and the theoretical line of thought.

```
n=1
Constant 0. Result: constant.
Constant 1. Result: constant.
Balanced. Result: balanced.
Balanced opposite. Result: balanced.

n=6
Constant 0. Result: constant.
Constant 1. Result: constant.
Balanced. Result: balanced.
Balanced opposite. Result: balanced.
```
Listing 7.4 Deutsch–Jozsa algorithm Q# output.

7.2   Bernstein–Vazirani Algorithm

The Bernstein–Vazirani problem builds on the work done by Deutsch and Jozsa, and was introduced by Ethan Bernstein and Umesh Vazirani [3]. It relies on the exact some circuit design for the algorithm as the Deutsch–Jozsa circuit we pictured in Fig. 7.3, but it defines a different problem that this oracle-based circuit is capable of answering.

7.2.1   Bernstein–Vazirani Problem

The problem is formulated as follows. Given a black-box function

$$f : \{0, 1\}^n \to \{0, 1\} \tag{7.26}$$

such that $f(x) = s \cdot x$, where s is a constant vector of $\{0, 1\}^n$ and $s \cdot x$ is a scalar product, find the bits making up s. In other words, the function takes an n bit input, and produces a single bit output, which is a bitwise product of the input bits with unknown bit array s, and the task is to find that s.

Classically, the solution to this problem is rather tedious, as it requires n steps. For example for a four bit input, we would need to query the function with inputs of 1000, 0100, 0010 and 0001, each of which would help us determine a different bit of s. Merging of those results will then uncover the value of s. On the other hand, using a quantum computer, the value of s can be found in a single logical query.

Because the quantum circuit for Bernstein–Vazirani algorithm is identical to the generalized Deutsch–Jozsa algorithm, almost all of the theoretical reasoning from Sect. 7.1 can be reused. We can begin here using the Deutsch–Jozsa state $|\psi_4\rangle$ from Eq. (7.21), which describes the overall system state after the $\mathbf{U_f}$ oracle application. This is the state that exhibits the phase kickback behavior, and it is integral to the success of the Bernstein–Vazirani algorithm too. To avoid collisions with the equations from Sect. 7.1, we will use $|\varphi_n\rangle$ for state notation here.

$$|\varphi_1\rangle = \frac{1}{\sqrt{2^n}} \sum_{x=0}^{2^n-1} (-1)^{f(x)} |x\rangle |-\rangle \tag{7.27}$$

At this point we know already that the trailing auxiliary qubit can be safely dropped from any consideration as it will stay in state $|-\rangle$. Additionally, in the Bernstein–Vazirani problem statement, $f(x) = s \cdot x$. Taking both of these into account, we can simplify the state that is interesting to us into $|\varphi_2\rangle$:

$$|\varphi_2\rangle = \frac{1}{\sqrt{2^n}} \sum_{x=0}^{2^n-1} (-1)^{s \cdot x} |x\rangle \tag{7.28}$$

Let us now expand part of the state $|\varphi_2\rangle$ for readability.

$$\sum_{x=0}^{2^n-1} (-1)^{s \cdot x} |x\rangle = \tag{7.29}$$

$$(|0\rangle + (-1)^{s_1} |1\rangle) \otimes (|0\rangle + (-1)^{s_2} |1\rangle) \otimes \cdots \otimes (|0\rangle + (-1)^{s_n} |1\rangle)$$

Looking at it from that angle, something really important stands out. Namely, if a given qubit is in a state $|+\rangle$, then its corresponding s_n must be equal to 0, because $(-1)^0 = 1$, so the term $(|0\rangle + (-1)^{s_1} |1\rangle)$ resolves to $(|0\rangle + |1\rangle)$. Conversely, when the qubit is in state $|-\rangle$, then s_n must be equal to 1.

Therefore, all we should be interested in at this point to perform an X-basis measurement—something that can be achieved by applying the $\mathbf{H}^{\otimes n}$ to state $|\varphi_2\rangle$, followed by the measurement in the Z-basis. The $\mathbf{H}^{\otimes n}$ transformation produces state equal to the one in Eq. (7.24), but that is effectively equal to $|s\rangle$, so measuring the input qubits now reveals the value of s in a single logical function evaluation.

Fig. 7.8 Bernstein–Vazirani
algorithm $\mathbf{U_f}$
implementation for $s = 101$

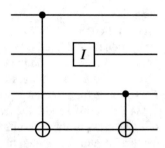

The quantum circuit representing the inner workings of the oracle, to satisfy
the requirements of the Bernstein–Vazirani algorithm, is going to be similar to the
balanced case for Deutsch–Jozsa algorithm from Fig. 7.6. In this case we will have a
dot product between the auxiliary qubit and the relevant nth qubit, so a set of **CNOT**
gates, applied only where the secret has bit value 1. A specific example for a secret
equal to 101 is shown in Fig. 7.8.

7.2.2   Q# Implementation

At this point we have all the theory needed to implement a Q# version of Bernstein–
Vazirani algorithm. For reasons that will become apparent in a moment, we will
define the initial implementation in a single operation only. To be more precise, we
will not treat the oracle as a separate operation, like we did that in Deutsch–Jozsa Q#
code, but rather we will inline the oracle implementation into the rest of the algorithm
code. This of course is hardly corresponding to a real life application, and is not a
best programming practice, though it will allow us to keep things simple for now.

The code is shown in Listing 7.5 and follows the general Deutsch–Jozsa circuit
from Fig. 7.3, enriched with the inlined oracle described by Fig. 7.8. This specific
oracle circuit was an example for the case where $s = 101$, however using a simple
loop, we can however generalize this and facilitate any secret within an arbitrarily
chosen bound. The chosen bound is three bits, and the algorithm can be tested for
all of them—the significance of that is that there are $2^3 = 8$ secrets to be verified.
What is worth mentioning is that the amount of qubits in the main registry $|x\rangle$ must
be large enough to hold the secret s. In this example this is three (qubits, and bits,
respectively)—however, if the secret cannot be encoded into the allocated qubits, the
algorithm would not work.

```
@EntryPoint()
operation Main() : Unit {
    let bitSize = 3;
    for intSecret in 0 .. 7 {
        let secret = IntAsBoolArray(intSecret, bitSize);
        use (x, y) = (Qubit[bitSize], Qubit());
        X(y);
        ApplyToEach(H, x);
```

```
H(y);

// oracle
for i in 0 .. Length(x) - 1 {
    if secret[i] {
        CNOT(x[i], y);
    }
}

ApplyToEach(H, x);
let result = MultiM(x);
Reset(y);
let intResult = ResultArrayAsInt(result);
Message($"Expected secret: {intSecret}, found secret: {
    intResult}");
    }
}
```

Listing 7.5 Simple implementation of Bernstein–Vazirani algorithm in Q#, with inlined oracle code.

A secret is represented by a boolean array obtained by converting the integer iteration index, 0–7, using the standard Q# `IntAsBoolArray` (number : Int, bits : Int) conversion function from the `Microsoft.Quantum.Convert` namespace. The oracle part applies the **CNOT** between the given qubit and the auxiliary qubit when the corresponding secret bit is 1. Finally, the secret is read out by performing a measurement over the qubits making up the $|x\rangle$ part of the system—the input register. The measurement is done with `MultiM` (targets : Qubit[]) operation, same as we did this in Deutsch–Jozsa algorithm implementation.

The final part of the code outputs the read out secret, compared to the expected secret—naturally, they should be equal. Since the source for computing the secret was the iteration index integer, the measurement results are converted into an integer as well using another function from the Q# standard library, `ResultArrayAsInt` (results : Result[]). This makes it much simpler to verify the outcome of the protocol—and the output of this code should be the same as in Listing 7.6.

```
Expected secret: 0, found secret: 0
Expected secret: 1, found secret: 1
Expected secret: 2, found secret: 2
Expected secret: 3, found secret: 3
Expected secret: 4, found secret: 4
Expected secret: 5, found secret: 5
Expected secret: 6, found secret: 6
Expected secret: 7, found secret: 7
```

Listing 7.6 Output of the basic implementation of Bernstein–Vazirani algorithm in Q#.

Of course such structuring (or lack thereof) of code is not particularly elegant, so it would make sense to rewrite it a program where the oracle is a standalone operation. Such extracted oracle, where the inputs are the secret, as `Bool[]`, the main qubit register x and the auxiliary qubit y, is shown in Listing 7.7. Notice that because the secret is an input into the oracle, we can avoid explicitly hardcoding it into the internals of the oracle implementation.

```
operation Oracle(secret : Bool[], x : Qubit[], y : Qubit) : Unit
    is Adj + Ctl  {
    for i in 0 .. Length(x) - 1 {
        if secret[i] {
            CNOT(x[i], y);
        }
    }
}
```

Listing 7.7 Bernstein–Vazirani algorithm implementation in Q#—a standalone oracle.

The algorithm itself will be now moved from the program's entry point into its own dedicated operation, similar to what we set up for Deutsch–Jozsa in Listing 7.2. As input it will take the expected registry size n, which will decide how many qubits to allocate in to the main register, and the reference to the oracle. This allows the caller to pass in various variants of the oracle, which is very helpful in testing the algorithm against different implementations of the U_f.

```
operation BernsteinVazirani(n : Int,
        oracle : ((Qubit[], Qubit) => Unit is Adj + Ctl))
        : Result[] {
    use (x, y) = (Qubit[n], Qubit());
    X(y);
    ApplyToEach(H, x);
    H(y);

    oracle(x, y);

    ApplyToEach(H, x);

    let result = MultiM(x);
    Reset(y);
    return result;
}
```

Listing 7.8 Bernstein–Vazirani algorithm implementation in Q#, with externally provided oracle.

The next natural step would be to bring the two last code chunks together using an orchestrator, and invoke the algorithm for various secret compositions. However, this is where things get slightly more challenging—the signatures of our two operations are not compatible with each other. The oracle's input arguments are (secret : Bool[], x : Qubit[], y : Qubit), while the algorithm expects the oracle as (Qubit[], Qubit) => Unit, so without the secret parameter. The solution is to create a third operation which will use Q#'s partial function application feature, and will allow us to prepare the oracle in the desired state. As a result, we are still able to parameterize the secret, but we can also pass the oracle into the algorithm's implementation as if the secret was hardcoded in the first place.

```
function PrepareOracle(secret : Bool[]) : ((Qubit[],Qubit) =>
        Unit is Adj + Ctl) {
    return Oracle(secret, _, _);
}
```

Listing 7.9 Bernstein–Vazirani algorithm implementation in Q#, oracle preparation.

The final step is to add the orchestration code, that will use similar iterative logic to test all secrets between 0 and 7, and which will glue all of the above pieces together. The output of this revised version of the Bernstein–Vazirani algorithm implementation will be the same as in Listing 7.6.

```
@EntryPoint()
operation Main() : Unit {
    let bitSize = 3;
    for intSecret in 0 .. 7 {
        let secret = IntAsBoolArray(intSecret, bitSize);
        let oracle = PrepareOracle(secret);
        let result = BernsteinVazirani(bitSize, oracle);
        let intResult = ResultArrayAsInt(result);
        Message($"Expected secret: {intSecret}, found secret: {
            intResult}");
    }
}
```

Listing 7.10 Revised implementation of Bernstein–Vazirani algorithm in Q#.

7.3 Grover's Algorithm

Grover's algorithm [17] was developed in 1996 by Indian-American computer scientist Lov Grover[1] who, using some particularly innovative algebraic techniques, found the first efficient quantum informational algorithm for searching through an unstructured data set.

We can state the problem for Grover's search algorithm in the following way. Given a black-box function

$$f : \{0, 1\}^n \to \{0, 1\} \tag{7.30}$$

find the single input $x \in \{0, 1\}^n$ such that $f(x) = 1$.

In other words, the Grover's algorithm requires an oracle able to determine whether a given sequence of bits is the result of the search, something that the oracle conveys using a single bit output 1 (for a match) and 0 for non-match. Of course such formulation generalizes the algorithm very well, and in such way it can now be applied to a very wide array set of problems—the black-box is simply fed possible results, and tests whether the given input is the solution. A subtle but important thing to remark here—as Bera [2] points out, the oracle does not need to *know* the answer, it only needs to be able to *recognize* it.

The first physical implementation of Grover's algorithm was achieved by Chuang et al. [5], using their two-qubit nuclear magnetic resonance quantum computer.

---

[1] Interestingly, it was one of the examples of a major contribution to the field of quantum computing made from outside of academia. Grover was long affiliated with Cornell University, however at the time the algorithm was developed, he worked exclusively at Bell Labs.

Classically, for N elements, a search through an unsorted data set has a complexity of $O(N)$. Grover showed, that a quantum computer can solve the same problem with the algorithmic complexity $O(\sqrt{N})$, resulting in quadratic speed up. In terms of raw steps to be evaluated, a classical brute-force approach,[2] where every item needs to be tested to verify if it is possibly the solution, would yield the answer in an average of $\frac{N}{2}$ checks. In worst case scenario, the answer may even come in $N-1$ queries. In Grover's algorithm, the logical step count for a quantum computer is approximately $\frac{\pi}{4}\sqrt{N}$.

While, in terms of complexity theory, quadratic speed-up may not offer a dramatic performance boost, it is still quite significant. Consider attempting to find a single specific 32-bit unsigned integer. There are $N = 2^{32} = 4\,294\,967\,295$ of them, so a classical algorithm requires on average:

$$\frac{4\,294\,967\,295}{2} \approx 2^{31} = 2\,147\,483\,648 \tag{7.31}$$

steps. On the other hand, Grover's algorithm needs far fewer steps:

$$\frac{\pi}{4}\sqrt{4\,294\,967\,295} = \frac{\pi}{4} \times 2^{16} \approx 51\,472 \tag{7.32}$$

In this account, we shall assume that we work with data sets whose length is proportional powers of two—namely that $N = 2^n$. This will make it easy to scale our reasoning process—for example, for three qubits, the data set will be able to accommodate eight possible data points, for four it would be sixteen, for five twenty-five and so on. In reality for data that does not correspond to powers of two, some padding would be required.

The original paper from Grover [17] used a phone book example—finding the person by their phone number. Other sources have often referred to Grover's algorithm as *database search*. While these analogies work well from illustrative standpoints, they are hardly practical in terms of real world applications, given the memory and I/O limitations of current quantum hardware. Not only would it be difficult to fit the working dataset in them, as machines with large, stable qubit counts are not available, the mere fact that the data would need to be preloaded into quantum computer in the first place has the $O(N)$ complexity already.

However, as Mermin [21] points out, Grover's algorithm can potentially be very useful to work on problems where the data set would be created in the quantum computer during the algorithm's execution, such as for example various mathematical problems or physical simulations.

---

[2] For the sake of simplicity we are going to assume the classical method for searching through unsorted dataset requires traversing the entire space of possible solutions. This is not always the case, as there can be classical probabilistic algorithms, which, for example, can look for most plausible solution candidates first, or which may exclude entire groups of answers.

7.3.1 Grover's Oracle

When implementing Grover's algorithm, one can rely on two types of oracle functions. The first one, often referred to as *boolean oracle*, relies on flipping the sign of the auxiliary qubit to hold the answer. Such oracles are typically denoted as U_f in literature, and ultimately need to rely on a similar phase kickback trick as Deutsch–Jozsa algorithm, which is facilitated by the auxiliary qubit being in state $|-\rangle$. The second type of oracles are the *phase oracles*, which flip the phase of the qubits using controlled Z gates. Because the phase in the main register $|x\rangle$ is manipulated directly, no additional auxiliary qubits are needed. This class of oracles is commonly denoted as U_ω, where ω corresponds the solution that is being searched for. The two oracle types are often used synonymously which can be confusing—in this study, we shall explicitly use the U_ω phase oracle.

Overall, we can describe the unitary transformation U_ω as:

$$U_\omega |x\rangle = \begin{cases} |x\rangle & \text{if } f(x) = 0 \\ -|x\rangle & \text{if } f(x) = 1 \end{cases} = \begin{cases} |x\rangle & \text{if } x \neq \omega = 0 \\ -|x\rangle & \text{if } x = \omega = 1 \end{cases} \tag{7.33}$$

where $|x\rangle$ can encompass n qubits. We can rewrite it into a more concise, generic, definition for the oracle as:

$$U_\omega |x\rangle = (-1)^{f(x)} |x\rangle \tag{7.34}$$

Let us illustrate this with a practical example. Consider a data set of size $N = 8$, spanning across three-qubit system, in a uniform superposition:

$$\frac{1}{\sqrt{8}} \left(|000\rangle + |001\rangle + |010\rangle + |011\rangle + |100\rangle + |101\rangle + |110\rangle + |111\rangle \right) \tag{7.35}$$

We can plot these probability amplitudes on a diagram. Each of the possible four measurement outcomes has an associated probability amplitude of $\frac{1}{\sqrt{8}}$ and, according to the Born rule, a corresponding classical probability of 12.5%. This is shown in Fig. 7.9.

Now let us now imagine that our oracle U_ω flags $|100\rangle$ as the correct answer $|\omega\rangle$, thus flipping the sign of the associated amplitude, and leaving the others unchanged. This leads us to state

$$\frac{1}{\sqrt{8}} \left(|000\rangle + |001\rangle + |010\rangle + |011\rangle - |100\rangle + |101\rangle + |110\rangle + |111\rangle \right) \tag{7.36}$$

Now each of the eight possible measurement results still has the classical probability of 12.5%, but the expected solution state $|\omega\rangle = |100\rangle$ has the associated probability amplitude that is different from the other potential outcome states. This effect of the oracle is shown in Fig. 7.10.

Fig. 7.9 Probability
amplitudes of the eight
possible states between $|000\rangle$
and $|111\rangle$ when the system is
in a uniform superposition
state

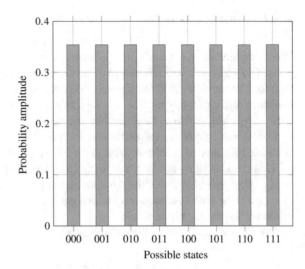

Fig. 7.10 Probability
amplitudes when the system
is in a uniform superposition
state, after the state $|100\rangle$
received a phase flip

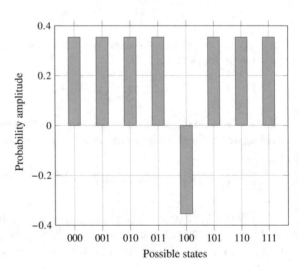

As we already know, in order to achieve any computational advantage over classical computing, the input into the oracle should already be the superposition of all
possible states. This of course makes sense from another perspective too—since at
the beginning of the algorithm, we have no idea where specifically the answer $|\omega\rangle$
is located, a uniform superposition with each possibility given equal weight, seems
like the best logical place to start. Therefore, the initial step in Grover's algorithm
will be the $\mathbf{H}^{\otimes n}$ transformation.

$$|\psi_1\rangle = \mathbf{H}^{\otimes n} |x\rangle \tag{7.37}$$

With a data set of N elements, this can be expanded into

$$|\psi_1\rangle = \frac{1}{\sqrt{N}} \sum_{x=0}^{N-1} |x\rangle \qquad (7.38)$$

The next step will be to query the oracle $\mathbf{U}_\omega$, passing in $|\psi_1\rangle$ as the input state. The oracle will shift the phase of the solution according to the rules we outlined in Eq. (7.33). This leads to a general system state:

$$|\psi_2\rangle = \mathbf{U}_\omega |\psi_1\rangle = \frac{1}{\sqrt{N}} \sum_{x=0}^{N-1} (-1)^{f(x)} |x\rangle \qquad (7.39)$$

Because the phase shift is applied only to the solution $|\omega\rangle$, we can rewrite the state $|\psi_2\rangle$ to better illustrate the effect of the oracle.

$$|\psi_2\rangle = -\frac{1}{\sqrt{N}} |\omega\rangle + \frac{1}{\sqrt{N}} \sum_{\substack{x \in \{0,1\}^n \\ x \neq \omega}} |x\rangle \qquad (7.40)$$

Such reformulation emphasizes that we now moved from a state of equal super-position, with each possible outcome described by the same positive amplitude, to a state where the correct answer state $|\omega\rangle$ now has a *negative* amplitude, while all the other incorrect answers states remain with the positive amplitudes—exactly how we visualized this in Fig. 7.10. While this appears quite promising, phase difference are in principle not observable, because the classical probabilities still remain the same. However, the brilliant innovation of Grover was that he realized that there a exists a general mechanism that allows to convert phase difference into amplitude difference, thus impacting the probability of measuring the correct result.

7.3.2 Inversion About the Mean

Grover used a mathematical tool called inversion about the mean to achieve this. Let us consider the example from Fig. 7.9 again. In that case, we dealt with eight possible measurement results, each one described by a probability amplitude $\frac{1}{\sqrt{8}}$. Since they are all equal to each other, their mean value m is also the same:

$$m = \frac{1}{\sqrt{8}} \approx 0.354 \qquad (7.41)$$

However, as soon as the oracle $\mathbf{U}_\omega$ acts on the system, and the state $|\omega\rangle$ acquires a phase sign different from the other possible measurement outcomes, the mean value m of the amplitudes is different.

$$m = \frac{6 \times \frac{1}{\sqrt{8}}}{8} = \frac{3\sqrt{2}}{16} \approx 0.265 \tag{7.42}$$

All values v in the data set can be inverted about the mean m of that data set. The procedure does not change the mean value, the sum of all components of the data set remains the same, and the relative distance to the mean is also still the same for each of the data points. Inversion about the mean, however, does allow to increase the distance between the selected value and the other values in the data set, making it *stand out* more than before.

The formula to obtain the inverted result v' is

$$v' = m + (m - v) = 2m - v \tag{7.43}$$

If we apply the formula to the mean we just computed in Eq. (7.42) and use amplitude $v = -\frac{1}{\sqrt{8}}$ corresponding to answer state $|\omega\rangle = |100\rangle$, we arrive at

$$v' = \frac{2 \times 3\sqrt{2}}{16} + \frac{1}{\sqrt{8}} = \frac{5\sqrt{2}}{8} \approx 0.884 \tag{7.44}$$

This boosted the classical probability for reading out the correct result from $\approx 0.354^2 = 12.5\%$ to $\approx 0.884^2 \approx 78.1\%$! Additionally, each of the remaining amplitudes—all of them equal to $v = \frac{1}{\sqrt{8}}$, need to go through the same process as well, to ensure the mean remains the same. Since $|\omega\rangle = |100\rangle$ experienced a drastic increase in the amplitude value, all other amplitudes should be decreased proportionally.

$$v' = \frac{2 \times 3\sqrt{2}}{16} - \frac{1}{\sqrt{8}} = \frac{\sqrt{2}}{8} \approx 0.177 \tag{7.45}$$

Overall, Fig. 7.11 shows the effect of the inversion about the mean on the probability amplitude distribution in a three qubit system, given the initial distribution equal to that which was defined in Fig. 7.10.

While this makes sense logically, we need to find the appropriate algebraic mechanism to achieve the same result when working with quantum states in a quantum computer. Thankfully, the same inversion about the mean procedure can be applied to vector states, with the obvious change that the formula would no longer operate on numbers but instead would have a matrix representation and would perform vector transformations. The generic vector-based formula would be

$$\mathbf{V}' = \mathbf{V}(2\mathbf{M} - \mathbf{I}) \tag{7.46}$$

where $\mathbf{M}$ is the matrix that finds the mean of the vector's components, $\mathbf{I}$ is the identity matrix and $\mathbf{V}$ and $\mathbf{V}'$ are multi dimensional vectors. Equipped with this knowledge, we are well positioned to find the next steps of the Grover's algorithm.

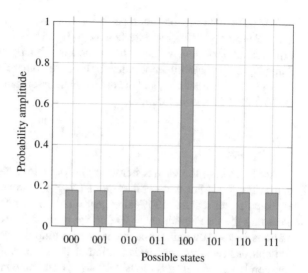

Fig. 7.11 Probability amplitudes after inversion about the mean amplified the amplitude for $|100\rangle$ and proportionally lowered all the other ones

7.3.3 Grover Diffusion Operator

Inversion about the mean procedure in a quantum algorithm can be generally realized using a conditional phase shift outlined in Eq. (7.47), applied against the balanced superposition of states.

$$|x\rangle \rightarrow \begin{cases} |x\rangle & \text{when } x = 0^n \\ -|x\rangle & \text{otherwise} \end{cases} \tag{7.47}$$

In other words, we would want every computational state except for $|00\dots0\rangle$ to receive a phase shift of -1. We can use that, as well as the prescription from Eq. (7.46) to define the inversion about the mean operator $\mathbf{U}_\psi$ as

$$\mathbf{U}_\psi = \mathbf{H}^{\otimes n}(2\,|0\rangle^{\otimes n}\,\langle0|^{\otimes n} - \mathbf{I}^{\otimes n})\mathbf{H}^{\otimes n} \tag{7.48}$$

Notice that this works because $\mathbf{H}$ is its own inverse, so surrounding $2\,|0\rangle^{\otimes n}\,\langle0|^{\otimes n} - \mathbf{I}$ with two $\mathbf{H}^{\otimes n}$ effectively results in un-computing and computing of a superposition. The above equation can be further simplified into

$$\mathbf{U}_\psi = (2\,|\psi\rangle\,\langle\psi| - \mathbf{I}^{\otimes n}) \tag{7.49}$$

which forms the final inversion about the mean operator, often referred to as *Grover diffusion operator*. The general matrix form of $\mathbf{U}_\psi$, for N possible states is:

$$\mathbf{U}_\psi = \begin{bmatrix} \frac{2}{N}-1 & \frac{2}{N} & \cdots & \frac{2}{N} \\ \frac{2}{N} & \frac{2}{N}-1 & \cdots & \frac{2}{N} \\ \vdots & \vdots & \ddots & \vdots \\ \frac{2}{N} & \frac{2}{N} & \cdots & \frac{2}{N}-1 \end{bmatrix} \tag{7.50}$$

The diffusion transformation performs what we call an *amplitude amplification*, and it relies on constructive interference to boost the probability amplitudes for the expected solution $|\omega\rangle$, while at the same time, thanks to destructive interference, reducing the probability amplitudes for all of the other invalid answers.

Together, the oracle $\mathbf{U}_\omega$ and the diffusion transformation $\mathbf{U}_\psi$ combine to make up the so-called *Grover iteration* $\mathbf{G}$:

$$\mathbf{G} = \mathbf{U}_\psi \mathbf{U}_\omega \tag{7.51}$$

Application of the Grover iteration allows us to achieve the effect we sketched out on Fig. 7.11. For two qubit systems ($N = 4$), the probability of measuring the expected result reaches 100% after just one Grover iteration. For larger data sets, including the three qubit ($N = 8$) one we used as an example, further G transformations, beyond the initial one, are needed to amplify the probabilities. The amount of iterations required for maximization of the probability amplitude of the correct solution is R, and depends on data set size N and amount of possible solutions M.

$$R \le \frac{\pi}{4}\sqrt{\frac{N}{M}} \tag{7.52}$$

Grover iterations are effectively rotations in the space spanned by the system state vector $|\psi\rangle$ and the state of the solution to the problem $|\omega\rangle$ (or uniform superposition of solution states, in case there are more than one). Because of that, past the R point, the usefulness of Grover iterations start to deteriorate rapidly, as the phase difference get smaller while amplitude differences are getting larger and larger. In other words, subsequent G transformations start to cause the probability amplitudes to drop, until they reach state in which the probability of reading out the correct answer is close to zero. From that point on, the rotations start to approach the solution state again. This cyclical, oscillating nature of Grover iteration, using $N = 16$ is shown in Fig. 7.12.

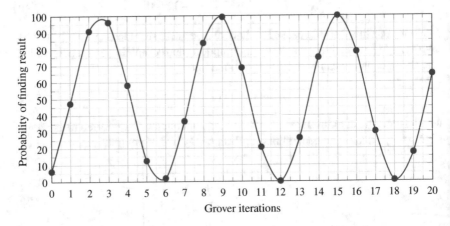

Fig. 7.12 Probability of reading the result correctly from a four qubit ($N = 16$) data set, after specific number of Grover iterations

It is important to remember that, all things considered, Grover's algorithm is still probabilistic, and even though the probabilities can be very high, it does not guarantee the deterministic results. The formula from Eq. (7.52) can help us determine when to stop the iterations within a single (first) oscillation—it is possible to reach higher correct answer probabilities on some of the subsequent oscillations, though that increases complexity of the search and negates the possible quadratic speed up.

The repeated Grover iterations mark the final part of Grover's algorithm. If we apply the $\mathbf{G}$ transformations R times to state $|\psi_2\rangle$ from Eq. (7.40), we will arrive at a final system state $|\psi_3\rangle$. Measurement of $|\psi_3\rangle$ would collapse to the desired state $|\omega\rangle$ with very high probability.

$$|\psi_3\rangle = (\mathbf{U}_\psi \mathbf{U}_\omega)^R |\psi_2\rangle = \mathbf{G}^R |\psi_2\rangle \approx |\omega\rangle \tag{7.53}$$

7.3.4 Review of Algorithm Steps

At this point we can review the overall recipe for Grover's algorithm, for problems with a single solution $M = 1$ (for problems with multiple solutions, the R bound would be different):

1. Prepare input state $|x\rangle = |0\rangle^{\otimes n}$
2. Apply $\mathbf{H}^{\otimes n}$ transformation to create a uniform superposition of all possible states
3. Apply Grover iteration $\mathbf{G}$ to the system $R \leq \frac{\pi}{4}\sqrt{N}$ times

   a. Apply the oracle $\mathbf{U}_\omega$
   b. Apply the inversion about mean transformation $\mathbf{U}_\psi$

4. Measure the input register $|x\rangle$ to obtain the result $|\omega\rangle$ with very high probability

A general quantum circuit for Grover's algorithm, with a single solution match, is shown in Fig. 7.13.

One thing we have not defined thus far are the internal implementations of both the oracle $\mathbf{U}_\omega$ and the inversion about mean transformation $\mathbf{U}_\psi$. Both of them can

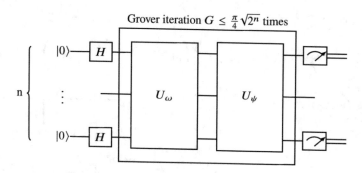

Fig. 7.13 Grover's algorithm quantum circuit

be implemented with the aid of the controlled phase shift gate **CZ**, which we first introduced in Sect. 4.6, spanning all n qubits.

Let us first cover the $\mathbf{U}_\omega$. The process for using the **CZ** gate to flip the sign on the solution is outlined below.

1. Convert the solution that is searched for into its binary representation. It should have the binary length as the size of the qubit register $|x\rangle$
2. Use the binary representation of the solution as a binary mask. For every solution jth bit that is equal to 0, apply the **X** transformation to the corresponding qubit in the register.
3. Apply **CZ** to flip the phase of the state $|11\ldots 1\rangle$
4. Apply the adjoint of the operation from step two. This is effectively undoing the step two transformation with the phase flip moving from the last term onto the solution

An example three qubit circuit following the procedure we just described, for the oracle $\mathbf{U}_\omega$ with $|\omega\rangle = |100\rangle$ as the solution, is shown in Fig. 7.14.

The diffusion transformation $\mathbf{U}_\psi$ can be realized via the following sequence of gates:

$$\mathbf{U}_\psi = \mathbf{H}^{\otimes n}\mathbf{X}^{\otimes n}\mathbf{CZ}^{\otimes n}\mathbf{X}^{\otimes n}\mathbf{H}^{\otimes n} \tag{7.54}$$

To better illustrate this, we can go through $\mathbf{U}_\psi$ step by step in a basic $n = 2$ qubit example. Such a simple two qubit system is very useful to study here, because already after a single amplitude amplification $\mathbf{U}_\psi$, the probability of reading out the solution flagged by the oracle increases to 100%. Consider the following uniform superposition state, where $|10\rangle$ was marked by the phase oracle $\mathbf{U}_\omega$ as the solution:

$$|\psi_1\rangle = \frac{1}{2}\Big(|00\rangle + |01\rangle - |10\rangle + |11\rangle\Big) \tag{7.55}$$

First, we apply $(\mathbf{H} \otimes \mathbf{H})$:

$$|\psi_2\rangle = (\mathbf{H} \otimes \mathbf{H})\,|\psi_1\rangle = \frac{1}{2}\Big(|00\rangle - |01\rangle + |10\rangle + |11\rangle\Big) \tag{7.56}$$

Followed by $(\mathbf{X} \otimes \mathbf{X})$:

$$|\psi_3\rangle = (\mathbf{X} \otimes \mathbf{X})\,|\psi_2\rangle = \frac{1}{2}\Big(|00\rangle + |01\rangle - |10\rangle + |11\rangle\Big) \tag{7.57}$$

Fig. 7.14 Sample circuit for $\mathbf{U}_\omega$ oracle, marking $|100\rangle$ as the answer

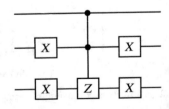

Then a two-qubit **CZ** flips the sign on the $|11\rangle$ term:

$$|\psi_4\rangle = \mathbf{CZ}\,|\psi_3\rangle = \frac{1}{2}\left(|00\rangle + |01\rangle - |10\rangle - |11\rangle\right) \tag{7.58}$$

Then we undo the earlier transformations by first applying $(\mathbf{X} \otimes \mathbf{X})$:

$$|\psi_5\rangle = (\mathbf{X} \otimes \mathbf{X})\,|\psi_4\rangle = \frac{1}{2}\left(-|00\rangle - |01\rangle + |10\rangle + |11\rangle\right) \tag{7.59}$$

And finally $(\mathbf{H} \otimes \mathbf{H})$ again, which leads us to the final state with a certain measurement outcome corresponding to the $|10\rangle$—and that is of course the answer that was initially indicated by the oracle.

$$|\psi_6\rangle = (\mathbf{H} \otimes \mathbf{H})\,|\psi_4\rangle = -|10\rangle \tag{7.60}$$

Naturally, as we already mentioned, this works so efficiently only for a two-qubit state, and for larger ones more Grover iterations would be needed, however it is still an elegant illustration of how the process of inverting about the mean works against quantum states. Notice that we started in Eq. (7.55) that had four probability amplitudes $\frac{1}{2}, \frac{1}{2}, -\frac{1}{2}$ and $\frac{1}{2}$ on all the four possible measurement outcomes respectively. The mean amplitude was therefore $\frac{1}{4}$. The Grover diffusion procedure resulted in all of the wrong terms, which were previously $\frac{1}{4}$ above the mean, to be decreased by $\frac{1}{2}$—so that they ended up $\frac{1}{4}$ below the mean. This resulted, very conveniently for us, for all of them receiving an amplitude 0. The correct term $|10\rangle$, on the other hand, was previously $\frac{3}{4}$ below the mean—flipping it $\frac{3}{4}$ above them made it reach the amplitude value 1 and thus probability of measurement equal to 1. Despite all of that, the mean amplitude remained unchanged at $\frac{1}{4}$.

To summarize, the amplitude amplification $\mathbf{U}_\psi$ corresponds to the example quantum circuit shown in Fig. 7.15, which is specifically constructed for a three-qubit case.

7.3.5 Q# Implementation

With all these pieces covered, we are well positioned to test out the logical behavior of Grover's algorithm by implementing it using Q# code. We will start by writing

Fig. 7.15 Circuit for executing amplitude amplification $\mathbf{U}_\psi$ in Grover's algorithm. Three qubits are shown but the pattern can be extended to any register size n

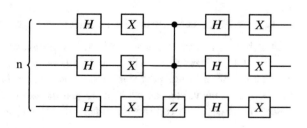

the code for the individual operations making up the Grover iteration—the oracle $\mathbf{U}_\omega$ and the diffusion transformation $\mathbf{U}_\psi$, as those are standalone, self-contained parts.

The code for the oracle is shown in Listing 7.11 and follows the sample circuit Fig. 7.14, with the major difference being that instead of being hardcoded to mark a specific answer, it can facilitate any predefined solution, of any size n. The core work of the oracle is of course the **CZ** gate spanning the entire qubit register. Because QDK does not provide a built-in n-qubit **CZ** operation, we can use the `Controlled` functor and the regular **Z** gate to achieve the same result. Additionally, because the oracle performs a set of conditional bitflips, and then performs the adjoint operation to undo those transformations, we can make use of Q# conjugation feature to make the code elegant and terse.

```
operation Oracle(solution : Int, qubits : Qubit[]) :
Unit is Adj {
let n = Length(qubits);
    within {
        let markerBits = IntAsBoolArray(solution, n);
        for i in 0..n-1 {
            if not markerBits[i] {
                X(qubits[i]);
            }
        }
    } apply {
        Controlled Z(Most(qubits), Tail(qubits));
    }
}
```

Listing 7.11 Q# implementation of the oracle for Grover's algorithm.

The oracle makes use of a helper function `IntAsBoolArray` (number : Int, bits : Int) from the `Microsoft.Quantum.Convert` namespace to convert the solution that the oracle knows into its binary representation. Note that the returned array uses a little-endian format, so for example the number 4 would become [`False`, `False`, `True`] not [`True`, `False`, `False`]. Then the oracle uses this bit array as a mask to apply the **X** transformations where needed. For simplicity we assume that the bit size n of the solution is smaller or equal to the length of the qubit array.

In order to separate the step of preparing the oracle with the solution, from its invocation as part of the Grover's algorithm and the individual Grover iterations, we will use partial function application as shown in Listing 7.12.

```
function PrepareOracle(solution : Int) : ((Qubit[]) => Unit is
    Adj) {
    return Oracle(solution, _);
}
```

Listing 7.12 Partial application of Grover's oracle to set the desired solution.

Another small feature worth mentioning is the usage of the `Most<'T>(array : 'T[])` and `Tail<'T> (array : 'T[])` functions which help split the qubits register into two blocks, one being an array containing all but the last element, and the other being just the last element itself. Those are then passed to the `Controlled` functor to perform the multi-qubit **CZ** transformation.

The inversion about the mean operation can also benefit from the same within...apply conjugation pattern as the oracle. The Q# implementation, following the circuit from Fig. 7.15 is shown in Listing 7.13.

```
operation Diffusion(qubits : Qubit[]) : Unit is Adj {
    within {
        ApplyToEachA(H, qubits);
        ApplyToEachA(X, qubits);
    } apply {
        Controlled Z(Most(qubits), Tail(qubits));
    }
}
```

Listing 7.13 Q# implementation of the amplitude amplification procedure for Grover's algorithm.

With the oracle and diffusor operations available, we can now construct the operation that will execute the entire Grover's algorithm. This operation is shown in Listing 7.14. As input, it will take the n bit size of the problem space, so that we can allocate the relevant amount of qubits and determine the correct iteration limit, as well as the oracle prepared according to the partial application logic from Listing 7.12. The Grover iterations will be limited by the variable r, which is the Q# code equivalent to the formula from Eq. (7.52).

```
operation Grover(n : Int, oracle : (Qubit[]) => Unit is Adj) : Int {
    let r = Floor(PI() / 4.0 * Sqrt(IntAsDouble(2 ^ n)));

    use qubits = Qubit[n];
    ApplyToEach(H, qubits);

    // Grover iterations
    for x in 1 .. r {
        oracle(qubits);
        Diffusion(qubits);
    }

    let register = LittleEndian(qubits);
    let number = MeasureInteger(register);
    return number;
}
```

Listing 7.14 Grover's algorithm in Q#.

The code follows the Grover circuit from Fig. 7.13. To make it more readable, the measurement process wraps the qubits in the LittleEdian register and uses that to perform a measurement of a single integer value, instead of reading out individual classical bits from the qubits. This makes it much easier to analyze the results, because such approach is symmetric to how we prepared the oracle (in Listing 7.11), where the solution was input as an integer as well. The oracle itself is also passed into the algorithm as an argument, namely as a delegate with the (Qubit[]) => Unit is Adj) signature. This will allow for the algorithm to be invoked with various oracles, and thus to test the search process against different solutions.

The final part of the process is to define an orchestration operation that will do just that—execute the algorithm a number of times and output the results for us to see and verify. The orchestration code using $n = 8$ is shown in Listing 7.15.

```
@EntryPoint()
operation Main() : Unit {
    let bitSize = 8;
    for solution in 0..2^bitSize - 1 {
        let oracle = PrepareOracle(solution);
        let result = Grover(bitSize, oracle);
        Message($"Expected to find: {solution}");
        Message($"Found: {result}");
    }
}
```

Listing 7.15 Grover's algorithm orchestration code.

The orchestration uses the bit size n of the problem space to loop through all the possible results—in this case 0–255, and prepares a corresponding oracle. In each case, the Grover's algorithm is invoked, and the outcome of the search is printed out to the console. The output should be similar to the (trimmed) one shown in Listing 7.16, which should also act as validation of our solution.

```
Expected to find: 0
Found: 0
Expected to find: 1
Found: 1
Expected to find: 2
Found: 2
...
Expected to find: 253
Found: 253
Expected to find: 254
Found: 254
Expected to find: 255
Found: 255
```

Listing 7.16 Grover's algorithm sample output.

7.4   Quantum Fourier Transform

7.4.1   Roots of Unity

Let us begin this section by introducing the concept of roots in mathematics. In general, an nth root of a number k is a number z that, when multiplied by itself n times, yields k.

$$z^n = k \tag{7.61}$$

A classical example could be $k = 4$, which has two second ($n = 2$) roots—either $z = -2$ or $z = 2$.

Roots get particularly interesting, once we constraint k by only allowing $k = 1$, which allows us to work with the so-called *root of unity*. For roots of unity, the convention is to use ω in place of z, so we shall also switch to that notation from this point on. This is still seemingly rather unremarkable, because on the surface, the first root ($n = 1$) of unity is $\omega = 1$ and the second root ($n = 2$) of unity are only $\omega_0 = 1$ and $\omega_1 = -1$. However, from the third root onwards, and thanks to usage of complex numbers, things become a lot more entertaining.

For $n = 3$, there are three roots of unity:

$$\omega_0 = 1 = e^{\frac{0\pi i}{3}} \tag{7.62}$$

$$\omega_1 = \frac{-1 + i\sqrt{3}}{2} = e^{\frac{2\pi i}{3}}$$

$$\omega_2 = \frac{-1 - i\sqrt{3}}{2} = e^{\frac{4\pi i}{3}}$$

Consequently, for $n = 4$, we can find four roots of unity, and those are easy to do because the imaginary number itself i, also becomes a fourth root. We can also see a pattern start to emerge[3]:

$$\omega_0 = 1 = e^{\frac{0\pi i}{4}} \tag{7.63}$$

$$\omega_1 = i = e^{\frac{2\pi i}{4}}$$

$$\omega_2 = -1 = e^{\frac{4\pi i}{4}}$$

$$\omega_3 = -i = e^{\frac{6\pi i}{4}}$$

Overall, the general formula for the Nth root of unity that can be spotted here is:

$$\omega_k = e^{\frac{2k\pi i}{N}} \tag{7.64}$$

The roots of unity are a de facto rotations around a circle divided into N parts— each time we multiply a root of unity by itself, we get a rotation $\frac{1}{N}$ of the way around the circle.

We will be needing this in a moment.

7.4.2 Discrete Fourier Transform

Fourier Transform, which was originally developed by French mathematician Joseph Fourier, is a linear transformation function that can be used to convert the function's time-domain representation to its frequency-domain representation. It's discrete variant, *Discrete Fourier Transform*, has achieved extraordinary success as it allowed discovery and extraction of frequency data from all kinds of signals. Because of that,

---

[3] We intentionally do not reduce the fractions to ensure the pattern really is visible.

it has found a very wide spectrum of applications in fields ranging from mathematics, music and sound engineering, through spectroscopy, to digital image processing.

The mathematical definition of DFT is as follows. Given an input set f of N complex numbers, DFT maps it to another complex set h so that:

$$h_j = \frac{1}{\sqrt{N}} \sum_{k=0}^{N-1} f_k e^{\frac{2\pi ijk}{N}} \tag{7.65}$$

We can spot here that $e^{\frac{2\pi i}{N}}$ is actually the primitive root of unity from Eq. (7.64), a fact that will allow us to simplify the math a lot.

$$h_j = \frac{1}{\sqrt{N}} \sum_{k=0}^{N-1} f_k \omega^{jk} \tag{7.66}$$

Additionally, we can also model DFT as a mapping of a vector f to a vector h. If we represent the number sets in such vector format, we can express DFT as a $N \times N$ matrix $\mathbf{U_{DFT}}$.

$$\begin{bmatrix} h[0] \\ h[1] \\ h[2] \\ \vdots \\ h[N-1] \end{bmatrix} = \mathbf{U_{DFT}} \begin{bmatrix} f[0] \\ f[1] \\ f[2] \\ \vdots \\ f[N-1] \end{bmatrix} \tag{7.67}$$

Taking Eqs. (7.64) and (7.66) into account, we arrive at the matrix form of the DFT, which consists of a combination of powers of Nth roots of unity:

$$\mathbf{U_{DFT}} = \frac{1}{\sqrt{N}} \begin{bmatrix} \omega^0 & \omega^0 & \omega^0 & \omega^0 & \dots & \omega^0 \\ \omega^0 & \omega^1 & \omega^2 & \omega^4 & \dots & \omega^{(N-1)} \\ \omega^0 & \omega^2 & \omega^4 & \omega^6 & \dots & \omega^{2(N-1)} \\ \omega^0 & \omega^3 & \omega^6 & \omega^9 & \dots & \omega^{3(N-1)} \\ \vdots & \vdots & \vdots & \ddots & \vdots & \vdots \\ \omega^0 & \omega^{(N-1)} & \omega^{2(N-1)} & \omega^{3(N-1)} & \dots & \omega^{(N-1)(N-1)} \end{bmatrix} \tag{7.68}$$

The fastest algorithmic implementation of DFT is *Fast Fourier Transform* [7], which reduces the complexity of computing the classical DFT from $O(N^2)$ to $O(N log N)$, where N is the data size. The main limitation of FFT is that it is restricted for N signal sizes that are power-of-two, so data padding may be required. Regardless of this dramatic computational speed-up provided by FFT, as pointed out by Zygelman [28], the resources needed for computation still grow exponentially, if the input list grows exponentially. As a result, trying to compute DFT for all permutations of the binary input $N = 2^n$, one arrives at very rapid exhaustion of resources on classical computers.

7.4.3 Quantum Fourier Transform

Quantum Fourier Transform resembles closely the Digital Fourier Transform and provides an incredible—exponential—computational speed up over the Fast Fourier Transform. The initial version of the quantum variant of FFT was developed in 1994 in IBM by Peter Shor [23], as part of his work on the factoring algorithm. Soon after, another IBM mathematician, Don Coppersmith, developed an improved version that required $O(log^2 N)$ steps, thus achieving the exponential speedup over its classical counterpart [8]. The technique became a critical component of Shor's factoring algorithm and has since found other useful application scenarios in quantum computer science.

Let us see here how DFT can be modeled as quantum state transformation, so that we can apply it in a quantum program. Quantum Fourier Transform $\mathbf{U_{QFT}}$ is effectively identical transformation to Discrete Fourier Transform [22]. If we only consider basis states and ignore the possible superpositions for now, it can be described by rewriting Eq. (7.66) into braket notation and updating the sum to be done over 2^n, because that is the amount of possible states to be encoded into n qubits as shown in Eq. (7.69) (where $|x\rangle$ is a basis state).

$$\mathbf{U_{QFT}} |x\rangle = \frac{1}{\sqrt{2^n}} \sum_{y=0}^{2^n-1} \omega^{xy} |y\rangle \tag{7.69}$$

To extrapolate it onto general superposition state, we can recall that in general we can express an arbitrary n qubit superposition state $|\psi\rangle$ using the following formula:

$$|\psi\rangle = \sum_{j=0}^{2^n-1} x_j |j\rangle \tag{7.70}$$

In other words, the quantum state is characterized by 2^n complex-numbered probability amplitudes x_j. The sum of all amplitudes x_j must naturally be equal to one. We also know that an arbitrary quantum state transformation $\mathbf{U}$, acting on this generalized state $|\psi\rangle$, would need to produce a new superposition state, so it would map $|j\rangle$ to $|k\rangle$ and x_j to y_k

$$\mathbf{U} |\psi\rangle = \mathbf{U} \sum_{j=0}^{2^n-1} x_j |j\rangle = \sum_{k=0}^{2^n-1} y_k |k\rangle \tag{7.71}$$

Taking all of this into account, we can combine Eqs. (7.69) and (7.71) together such that the resulting Quantum Fourier Transform would enact state evolution where the amplitudes of the original input state are related to the amplitudes of the superposition output state by the Discrete Fourier Transform.

As a result, we arrive at the final definition of Quantum Fourier Transform, acting upon an arbitrary superposition state $|\psi\rangle$.

$$\mathbf{U_{QFT}}|\psi\rangle = \frac{1}{\sqrt{2^n}} \sum_{j=0}^{2^n-1} \sum_{k=0}^{2^n-1} x_k \omega^{jk} |j\rangle \tag{7.72}$$

It can also be calculated that

$$\mathbf{U_{QFT}} \mathbf{U_{QFT}}^\dagger = 1 \tag{7.73}$$

where $\mathbf{U_{QFT}}^\dagger$ is the Hermitian adjoint of $\mathbf{U_{QFT}}$. This of course means that the $\mathbf{U_{QFT}}$ transformation is unitary, and thus suitable for usage as a quantum state transformation during a computation process on a quantum hardware.

The matrix representation for the QFT transformation would follow that of the DFT one, defined in Eq. (7.68). With a few simplifications applied, it looks like this:

$$\mathbf{U_{QFT}} = \frac{1}{\sqrt{2^n}} \begin{bmatrix} 1 & 1 & 1 & 1 & \dots & 1 \\ 1 & \omega^1 & \omega^2 & \omega^4 & \dots & \omega^{(2^n-1)} \\ 1 & \omega^2 & \omega^4 & \omega^6 & \dots & \omega^{(2^n-2)} \\ 1 & \omega^3 & \omega^6 & \omega^9 & \dots & \omega^{(2^n-3)} \\ \vdots & \vdots & \vdots & \vdots & \ddots & \vdots \\ 1 & \omega^{(2^n-1)} & \omega^{(2^n-2)} & \omega^{(2^n-3)} & \dots & \omega \end{bmatrix} \tag{7.74}$$

This is probably still quite abstract, so what could help to illustrate QFT, would be to calculate the QFT matrices for one and two qubit cases. For a single qubit ($n = 1$) we deal with a tiny data set of size $N = 2^1 = 2$, and we can calculate the second root of unity using Eq. (7.64).

$$\omega = e^{\frac{2i\pi}{2}} = e^{i\pi} = -1 \tag{7.75}$$

We can now substitute that into the QFT matrix, and results in the following representation of a single qubit Quantum Fourier Transform:

$$\mathbf{U_{QFT_1}} = \frac{1}{\sqrt{2}} \begin{bmatrix} 1 & 1 \\ 1 & -1 \end{bmatrix} \tag{7.76}$$

Quite surprisingly, this ends up being exactly the structure of the $\mathbf{H}$ matrix, which tells us that for a single qubit Quantum Fourier Transform is the Hadamard transformation. This is, however, only a special case, and for larger qubit sizes $\mathbf{U_{QFT_n}} \neq \mathbf{H}^{\otimes n}$.

For $n = 2$, we arrive at

$$\omega = e^{\frac{2i\pi}{4}} = e^{\frac{i\pi}{2}} = i \tag{7.77}$$

which leads us the $\mathbf{U_{QFT_2}}$ matrix

$$\mathbf{U_{QFT_2}} = \frac{1}{2} \begin{bmatrix} 1 & 1 & 1 & 1 \\ 1 & i & -1 & -i \\ 1 & -1 & 1 & -1 \\ 1 & -i & -1 & i \end{bmatrix} \tag{7.78}$$

We can proceed to inspect the effects of this transformation on two qubit computational basis states $|00\rangle$, $|01\rangle$, $|10\rangle$ and $|11\rangle$.

$$\mathbf{U_{QFT_2}} |00\rangle = \frac{1}{2}(|00\rangle + |01\rangle + |10\rangle + |11\rangle) = \tag{7.79}$$

$$\frac{1}{\sqrt{2}}(|0\rangle + |1\rangle) \otimes \frac{1}{\sqrt{2}}(|0\rangle + |1\rangle)$$

$$\mathbf{U_{QFT_2}} |01\rangle = \frac{1}{2}(|00\rangle + i\,|01\rangle - |10\rangle - i\,|11\rangle) = \tag{7.80}$$

$$\frac{1}{\sqrt{2}}(|0\rangle - |1\rangle) \otimes \frac{1}{\sqrt{2}}(|0\rangle + i\,|1\rangle)$$

$$\mathbf{U_{QFT_2}} |10\rangle = \frac{1}{2}(|00\rangle - |01\rangle + |10\rangle - |11\rangle) = \tag{7.81}$$

$$\frac{1}{\sqrt{2}}(|0\rangle + |1\rangle) \otimes \frac{1}{\sqrt{2}}(|0\rangle - |1\rangle)$$

$$\mathbf{U_{QFT_2}} |11\rangle = \frac{1}{2}(|00\rangle - i\,|01\rangle - |10\rangle + i\,|11\rangle) = \tag{7.82}$$

$$\frac{1}{\sqrt{2}}(|0\rangle - |1\rangle) \otimes \frac{1}{\sqrt{2}}(|0\rangle - i\,|1\rangle)$$

A certain pattern should be visible in this—namely that $\mathbf{U_{QFT_2}}$ on two qubits produces a tensor product of single qubit superposition states, were the terms are equivalent to the Hadamard transformation. We can give it a more formal description be generalizing it onto any n number of qubits.

$$\mathbf{U_{QFT}} |j_1 \ldots j_{n-1} j_n\rangle = \tag{7.83}$$

$$\frac{1}{\sqrt{2^n}} \left(|0\rangle + e^{2\pi i \frac{j_n}{2^1}} |1\rangle\right) \otimes \left(|0\rangle + e^{2\pi i (\frac{j_{n-1}}{2^1} + \frac{j_n}{2^2})} |1\rangle\right) \otimes$$

$$\cdots \otimes \left(|0\rangle + e^{2\pi i (\frac{j_1}{2^1} + \cdots + \frac{j_{n-1}}{2^{n-1}} + \frac{j_n}{2^n})} |1\rangle\right)$$

The definition of $\mathbf{U_{QFT}}$ we managed to establish so far is helpful to algebraically illustrate its effects on quantum states, but cannot directly be applied as a single logic gate. We can, however, construct a circuit, made entirely out of one- and two-qubit gates, which will perform the Quantum Fourier Transform and be appropriate for usage on quantum hardware.

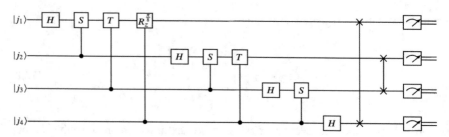

Fig. 7.16 QFT circuit for four qubits

The main building block of the circuit is the rotation gate $\mathbf{R}_z^{\varphi}$ which we introduced in Sect. 4.5, used in the controlled fashion over pairs of qubits, with rotation angles depending on the amount of qubits involved in the QFT process. We can also recall from that part of the book, that it is common to express $\mathbf{R}_z^{\varphi}$ with rotation angles of $\frac{\pi}{2}$ and $\frac{\pi}{4}$ as, $\mathbf{S}$ and $\mathbf{T}$ gates respectively.

The QFT circuit for four qubits is shown in Fig. 7.16.

Let us walk through the individual steps. We start off in state:

$$|j_1\rangle \otimes |j_2\rangle \otimes |j_3\rangle \otimes |j_4\rangle \tag{7.84}$$

After the initial $\mathbf{H}$ gate applied to the qubit $|j_1\rangle$, and using the mathematical formalism of $\mathbf{H}$ introduced in Eq. (4.13), the system state changes to:

$$\frac{1}{\sqrt{2}} \left(|0\rangle + e^{2\pi i \frac{j_1}{2^1}} \right) \otimes |j_2\rangle \otimes |j_3\rangle \otimes |j_4\rangle \tag{7.85}$$

Next, a controlled $\mathbf{S}$ is applied between the first and second qubits, leading us to state:

$$\frac{1}{\sqrt{2}} \left(|0\rangle + e^{2\pi i \left(\frac{j_1}{2^1} + \frac{j_2}{2^2} \right)} \right) \otimes |j_2\rangle \otimes |j_3\rangle \otimes |j_4\rangle \tag{7.86}$$

A controlled $\mathbf{T}$ gate follows, applying a further rotation to $|j_1\rangle$:

$$\frac{1}{\sqrt{2}} \left(|0\rangle + e^{2\pi i \left(\frac{j_1}{2^1} + \frac{j_2}{2^2} + \frac{j_3}{2^3} \right)} \right) \otimes |j_2\rangle \otimes |j_3\rangle \otimes |j_4\rangle \tag{7.87}$$

Finally, the controlled rotation $\mathbf{R}_z^{\frac{\pi}{8}}$ produces:

$$\frac{1}{\sqrt{2}} \left(|0\rangle + e^{2\pi i \left(\frac{j_1}{2^1} + \frac{j_2}{2^2} + \frac{j_3}{2^3} + \frac{j_4}{2^4} \right)} \right) \otimes |j_2\rangle \otimes |j_3\rangle \otimes |j_4\rangle \tag{7.88}$$

Similar procedure needs to be applied to the other qubits, each one receiving one less controlled rotation than the previous one. The second qubit will receive a gate sequence $\mathbf{H}$, controlled $\mathbf{S}$ and controlled $\mathbf{T}$, creating a system state:

$$\frac{1}{\sqrt{2}} \left(|0\rangle + e^{2\pi i \left(\frac{j_1}{2^1} + \frac{j_2}{2^2} + \frac{j_3}{2^3} + \frac{j_4}{2^4} \right)} \right) \otimes \tag{7.89}$$

$$\frac{1}{\sqrt{2}} \left(|0\rangle + e^{2\pi i (\frac{j_2}{2^1} + \frac{j_3}{2^2} + \frac{j_4}{2^3})} \right) \otimes$$
$$|j_3\rangle \otimes |j_4\rangle$$

Next, we apply the **H** and controlled **S** to the third qubit:

$$\frac{1}{\sqrt{2}} \left(|0\rangle + e^{2\pi i (\frac{j_1}{2^1} + \frac{j_2}{2^2} + \frac{j_3}{2^3} + \frac{j_4}{2^4})} \right) \otimes \tag{7.90}$$
$$\frac{1}{\sqrt{2}} \left(|0\rangle + e^{2\pi i (\frac{j_2}{2^1} + \frac{j_3}{2^2} + \frac{j_4}{2^3})} \right) \otimes$$
$$\frac{1}{\sqrt{2}} \left(|0\rangle + e^{2\pi i (\frac{j_3}{2^1} + \frac{j_4}{2^2})} \right) \otimes$$
$$|j_4\rangle$$

And finally, the last, fourth, qubit receives only the **H** transformation.

$$\frac{1}{\sqrt{2}} \left(|0\rangle + e^{2\pi i (\frac{j_1}{2^1} + \frac{j_2}{2^2} + \frac{j_3}{2^3} + \frac{j_4}{2^4})} \right) \otimes \tag{7.91}$$
$$\frac{1}{\sqrt{2}} \left(|0\rangle + e^{2\pi i (\frac{j_2}{2^1} + \frac{j_3}{2^2} + \frac{j_4}{2^3})} \right) \otimes$$
$$\frac{1}{\sqrt{2}} \left(|0\rangle + e^{2\pi i (\frac{j_3}{2^1} + \frac{j_4}{2^2})} \right) \otimes$$
$$\frac{1}{\sqrt{2}} \left(|0\rangle + e^{2\pi i \frac{j_4}{2^1}} \right)$$

This is a very promising result, because it is almost exactly the QFT output state which we defined in Eq. (7.83), with the only difference being the probability amplitudes, which are in the wrong order compared to what we were expecting. We can fix that using the **SWAP** gates—by swapping pairs of qubits $|j_1\rangle$ with $|j_4\rangle$ and $|j_2\rangle$ with $|j_3\rangle$. This is also the reason for including these swap operations in the circuit on Fig. 7.16. Note that the amount of swaps needed corresponds to the amount of qubits involved—at most, $\frac{n}{2}$ swaps will be necessary.

The final system state after the successful swapping of terms is:

$$\frac{1}{\sqrt{2}} \left(|0\rangle + e^{2\pi i \frac{j_4}{2^1}} \right) \tag{7.92}$$
$$\frac{1}{\sqrt{2}} \left(|0\rangle + e^{2\pi i (\frac{j_3}{2^1} + \frac{j_4}{2^2})} \right) \otimes$$
$$\frac{1}{\sqrt{2}} \left(|0\rangle + e^{2\pi i (\frac{j_2}{2^1} + \frac{j_3}{2^2} + \frac{j_4}{2^3})} \right) \otimes$$
$$\frac{1}{\sqrt{2}} \left(|0\rangle + e^{2\pi i (\frac{j_1}{2^1} + \frac{j_2}{2^2} + \frac{j_3}{2^3} + \frac{j_4}{2^4})} \right)$$

The final thing worth noting is that the approach discussed here assumes big endian sequencing of qubits. However, from programming perspective the opposite—little endian, which would simply use an inverted order of qubits—is a more natural and more commonly used method of working with bit registers in higher level languages. Q# provides built-in implementation for both big endian and little endian variants, and it is at the programmer's discretion to choose the method more appropriate for their use case.

7.4.4 Q# Implementation

The basic Q# implementation of a QFT is at this point not going to be particularly challenging, given that we have precise circuit definition. We will start with a
hardcoded four-qubit variant, mimicking Fig. 7.16.

In order to be able to see the effects of the QFT, we need to initialize the qubit
register to some non-zero value instead of executing it on the |0000⟩ state. To
achieve that, we use the helper operation SetValue (value : Int, qubits :
Qubit[]) shown in Listing 7.17.

```
operation SetValue(value : Int, qubits : Qubit[])
    : Unit is Adj + Ctl {
  ApplyToEachCA(
    CControlledCA(X),
    Zipped(IntAsBoolArray(value, Length(qubits)^2), qubits)
  );
}
```

Listing 7.17 Helper to initialize a qubit register to a specific integer value

The   helper   uses   the   built-in   Q#   core   operation
CControlled<'T> (op : ('T => Unit)),   which   allows   us   to
conditionally apply a given operation if the given classical bit is set to *true*.
In this case, we apply **X** which effectively results in the register being initialized to
the specific integer value.

The Q# code for a four qubit big endian QFT is shown next, in Listing 7.18.
Since the code is fixed to four qubits, we can just hardcode the register indices when
applying the relevant rotations, making the code very straight-forward. Notice that as
discussed in Sect. 4.5, we use R1(theta : Double, qubit : Qubit) instead of
Rz(theta : Double, qubit : Qubit) for rotation about the z-axis—this means
we are not changing the global phase. At the end, the operation uses a debug helper
function DumpMachine<'T> (location : 'T) which is useful to visualize the
effects of the QFT on the state of the system when running on the simulator. Additionally, since the qubits are not measured, they are manually reset. Obviously in real
world application these last two lines would be omitted.

```
operation FourQubitQFT_BE_Manual(initialValue : Int) : Unit {
    use qubits = Qubit[4];
    SetValue(initialValue, qubits);

    H(qubits[0]);
    Controlled S([qubits[1]], qubits[0]);
    Controlled T([qubits[2]], qubits[0]);
    Controlled R1([qubits[3]], (PI()/8.0, qubits[0]));

    H(qubits[1]);
    Controlled S([qubits[2]], qubits[1]);
    Controlled T([qubits[3]], qubits[1]);

    H(qubits[2]);
    Controlled S([qubits[3]], qubits[2]);
```

```
    H(qubits[3]);

    SWAP(qubits[2], qubits[1]);
    SWAP(qubits[3], qubits[0]);

    DumpMachine();
    ResetAll(qubits);
}
```

Listing 7.18 Manual QFT implementation in Q# for four qubits using big endian qubit register.

The hardcoded logic can relatively easily be generalized into a parameterized operation invoking the necessary amount of controlled R1 (theta
: Double, qubit : Qubit) rotation gates, with appropriate rotation angles,
and performing the required swaps depending on the size of the qubit register. In fact, Q# already contains an excellent built-in helper operation to
reverse the register order, SwapReverseRegister (register : Qubit[]) in the
Microsoft.Quantum.Canon namespace. The generalized approach for *n* qubits is
shown in Listing 7.19.

```
operation QFT_BigEndian_Manual(initialValue : Int) : Unit {
    use qubits = Qubit[BitSizeI(initialValue)];
    let length = Length(qubits);
    SetValue(initialValue, qubits);

    for i in 0..length-1 {
        H(qubits[i]);
        mutable power = 1;
        for j in i+1..length-1 {
            Controlled R1([qubits[j]], (PI() / PowD(2.0,
                IntAsDouble(power)), qubits[i]));
            set power += 1;
        }
    }

    SwapReverseRegister(qubits);

    DumpMachine();
    ResetAll(qubits);
}
```

Listing 7.19 Generalized QFT implementation in Q# for *n* qubits.

The size of the qubit register is dynamically computed based on the bit size of
the initial value set in the register. This is calculated using the Q# standard library
function BitSizeI (a : Int) from the Microsoft.Quantum.Math namespace.
The usages of **S** and **T** gates are now completely replaced with calls to R1 (theta :
Double, qubit : Qubit), with the rotation angles determined on-demand using
powers of two.

As nice as this code is to read and marvel, it is also equally unnecessary. That is
because, not surprisingly, Q# ships Quantum Fourier Transform implementations
out-of-the box. Therefore, in order to perform a big endian QFT in Q#, all we
need to do is wrap the qubit array into a BigEndian register and call QFT (qs

: BigEndian) operation from the `Microsoft.Quantum.Canon` namespace. This is shown in Listing 7.20.

```
operation QFT_BigEndian_Framework(initialValue : Int) : Unit {
    use qubits = Qubit[BitSizeI(initialValue)];
    SetinitialValue(initialValue, qubits);
    let register = BigEndian(qubits);

    QFT(register);

    DumpMachine();
    ResetAll(qubits);
}
```

Listing 7.20 Using the framework big endian QFT implementation in Q# for *n* qubits.

The result of the operations shown in Listings 7.19 and 7.20 is identical, with the obvious advantage that the latter is part of the framework and does not need to be written. If we were to execute any of them for a four qubit register, for example by initially setting the register to integer 10 ($|1010\rangle$), we would also get the effects identical to the operation listed in 7.18.

Finally, in case of working with the little endian register, we simply need to reverse the qubit order from Listing 7.18 when applying the controlled rotation sequences. Similarly as before, the manual implementation of the algorithm can be replaced with the framework-based one, `QFTLE (qs : LittleEndian)`. The additional change includes using a built-in operation `ApplyXorInPlace (value : Int, target : LittleEndian)` from the `Microsoft.Quantum.Arithmetic` namespace to initialize the register with a specific value. This has the same effect as our custom `SetValue(value : Int, qubits : Qubit[])` operation, but is readily available in the Q# standard library and works with little endian registers.

```
operation QFT_LittleEndian_Framework() : Unit {
    use qubits = Qubit[4];
    let register = LittleEndian(qubits);
    ApplyXorInPlace(8, register);

    QFTLE(register);

    DumpMachine();
    ResetAll(qubits);
}
```

Listing 7.21 Using the framework little endian QFT implementation in Q# for *n* qubits.

7.5 Quantum Phase Estimation

In Chap. 2 we introduced the linear algebra concepts of eigenvectors and eigenvalues. An eigenvector **x** of a matrix **A** is a vector that changes only by a scalar λ when the linear transformation represented by this matrix is applied to it. This has the general mathematical form of

$$\mathbf{A}\mathbf{x} = \lambda \mathbf{x} \tag{7.93}$$

Quantum phase estimation (QPE) is the eigenvalue problem in the context of quantum mechanics. All quantum state transformations must be unitary, so we would customarily denote them as $\mathbf{U}$, the length of the quantum state vector is always equal to 1 and, of course, it would typically be written in the braket notation. This leads us to the following way of expressing the eigenvalues of a unitary operators in quantum mechanics

$$\mathbf{U}\,|u\rangle = e^{2\pi i\theta}\,|u\rangle \tag{7.94}$$

where θ is the phase of the system, and takes values in the range of $0 \le \theta \le 1$. In other words, a unitary quantum state transformation $\mathbf{U}$ acting on its eigenvector $|u\rangle$, will produce the same quantum state vector $|u\rangle$ along with an eigenvalue $e^{2\pi i\theta}$, and the QPE algorithm allows us to estimate this phase θ for an arbitrary $\mathbf{U}$.

The problem has a wide spectrum of application in physics and technology, such as computing the ground state energy of a system or reducing the dimensionality of feature vectors in machine learning [18]. It is also an essential routine used as a building block in many other quantum algorithms such Shor's factoring algorithm or the quantum algorithm for linear systems of equations. Since the original QPE algorithm, which was developed by Kitaev [19], it has been revised and improved in a number of ways in terms of performance and circuit efficiency, for example by a group from Microsoft Research [25]. The implementation discussed here is a revised quantum phase estimation algorithm reported by Cleve et al. [6].

7.5.1 Core Theory

The core idea behind quantum phase estimation lies behind the fact that we assume we can express, with reasonable accuracy, the phase as n bits:

$$e^{2\pi i\theta} = e^{2\pi i 0.\theta_1\theta_2...\theta_n} \tag{7.95}$$

We can only speak about phase *estimation* because it is possible for θ to be a value that cannot be exactly represented with a fixed amount of bits—for example it could be an irrational value, such as $\frac{\sqrt{2}}{2}$. Because of this, when working with phase estimation algorithm, we will require to operate on two qubit registers, of sizes n and m respectively. The first register that will hold n qubits will determine the upper limit on the accuracy of the solution and the success probability for finding the phase correctly. The second register of size m and will accommodate the initial input state $|u\rangle$.

Another prerequisite for the algorithm is the assumption that we have access to a blackbox unitary operator $\mathbf{U}$, which is capable of both preparing the initial state $|u\rangle$ and executing controlled $\mathbf{U}^{2^j}$ operations for $0 \le j \le n - 1$.

A key thing to realize here is that we really need $\mathbf{U}^{2^j}$ because that is the way to reach the phase representation from Eq. (7.95).

$$\mathbf{U}^{2^j} \left|u\right\rangle = e^{2\pi i 2^j \theta} \left|u\right\rangle = e^{2\pi i 0.\theta_1 \theta_2 \ldots \theta_n} \left|u\right\rangle \tag{7.96}$$

A different way of looking at it is to consider what happens when $\mathbf{U}$ acts on its eigenvectors j amount of times [2]. To put it simply, a series of $\mathbf{U}$ ranging from $\mathbf{U}^{2^0}$ to $\mathbf{U}^{2^{j-1}}$ can be used to arrive at the correct value for the phase.

$$\mathbf{U}^{2^j} \left|u\right\rangle = \mathbf{U}^{2^j - 1} \mathbf{U} \left|u\right\rangle = \mathbf{U}^{2^j - 1} e^{2\pi i \theta} \left|u\right\rangle = \cdots = e^{2\pi i 2^j \theta} \left|u\right\rangle \tag{7.97}$$

These are exactly the ideas explored by the quantum phase estimation algorithm.

The circuit for quantum phase estimation algorithm is shown in Fig. 7.17. It is easiest to start exploring the algorithm by setting $n = 1$ and seeing how the system state changes.

$$\left|0\right\rangle \left|u\right\rangle \tag{7.98}$$

By placing the first register (in this case single qubit) in the superposition, we arrive at:

$$\frac{1}{\sqrt{2}} \left(\left|0\right\rangle + \left|1\right\rangle \right) \left|u\right\rangle = \frac{1}{\sqrt{2}} \left(\left|0\right\rangle \left|u\right\rangle + \left|1\right\rangle \left|u\right\rangle \right) \tag{7.99}$$

The controlled $\mathbf{U}$ then produces:

$$\mathbf{CU} \frac{1}{\sqrt{2}} \left(\left|0\right\rangle \left|u\right\rangle + \left|1\right\rangle \left|u\right\rangle \right) = \tag{7.100}$$

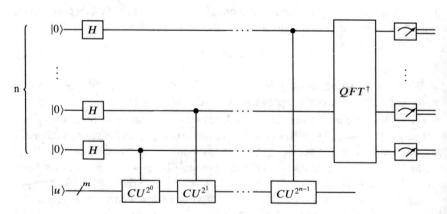

Fig. 7.17 QPE circuit

$$\frac{1}{\sqrt{2}} \left(|0\rangle |u\rangle + e^{2\pi i \theta} |1\rangle |u\rangle \right) = \frac{1}{\sqrt{2}} \left(|0\rangle + e^{2\pi i \theta} |1\rangle \right) |u\rangle$$

Because $|u\rangle$ is the eigenstate of $\mathbf{U}$ it is unaffected by it, and the phase information "spilled" into the first register and is now part of the single qubit state that we have there.

We are now well positioned to generalize together Eqs. (7.96), (7.97) and (7.100) to provide the formal algebraic description of the quantum phase estimation algorithm. The procedure starts with a system state $|\psi_1\rangle$:

$$|\psi_1\rangle = |0\rangle^{\otimes n} |u\rangle \tag{7.101}$$

This is followed by placing the first register in a uniform superposition by applying $\mathbf{H}^{\otimes n}$ to it.

$$|\psi_2\rangle = (\mathbf{H}^{\otimes n} \otimes \mathbf{I}^{\otimes m}) |\psi_1\rangle = \frac{1}{\sqrt{2^n}} \sum_{k=0}^{2^n-1} |k\rangle |u\rangle \tag{7.102}$$

Next, n controlled $\mathbf{U}^{2^j}$ operations are applied to $|\psi_2\rangle$. This leads, as we already saw in the simple example with a single qubit, to the encoding of phase information into the first register.

$$|\psi_3\rangle = \mathbf{U}^{2^j} |\psi_2\rangle = \frac{1}{\sqrt{2^n}} \left(|0\rangle + e^{2\pi i 2^{n-1} \theta} |1\rangle \right) \otimes \tag{7.103}$$

$$\left(|0\rangle + e^{2\pi i 2^{n-2} \theta} |1\rangle \right) \otimes \cdots \otimes \left(|0\rangle + e^{2\pi i 2^0 \theta} |1\rangle \right) \otimes |u\rangle$$

The state $|\psi_3\rangle$ can be written more concisely as:

$$|\psi_3\rangle = \frac{1}{\sqrt{2^n}} \sum_{k=0}^{2^n-1} e^{2\pi i \theta k} |k\rangle |u\rangle \tag{7.104}$$

At this point the second register become no longer needed, since it stays in state $|u\rangle$. Moreover, if we compare Eqs. (7.104) and (7.83) we will realize that the first register is now in a state that is the product of a Quantum Fourier Transform. Because in quantum mechanics all transformations are unitary, we can always apply an adjoint operation to revert a given state, which is exactly the algebraic trickery needed next.

By employing an adjoint of QFT, we are able to take the first register state from Eq. (7.104) to the state where the register actually contains an estimation of the phase we were looking for—$|\tilde{\theta}\rangle$.

$$\mathbf{U_{QFT}}^\dagger \left(\frac{1}{\sqrt{2^n}} \sum_{k=0}^{2^n-1} e^{2\pi i \theta k} |k\rangle \right) = |\theta_1\rangle \otimes |\theta_2\rangle \otimes \cdots \otimes |\theta_n\rangle = |\tilde{\theta}\rangle \tag{7.105}$$

A Z-basis measurement of the first register will now reveal to us the value of θ.

7.5.2   Q# Implementation

We can begin the Q# implementation of the algorithm by defining the oracle **U** and equipping it with a capability of being invoked as many times as needed, to correspond to the power-of-2 requirement that we have. A definition of such basic **U**, acting on a single qubit only, is provided in Listing 7.22. It could be used to wrap any single qubit unitary transformation, which is passed into the operation as the op argument. The other two arguments represent the power for which the oracle needs to be invoked and the qubits to act on.

```
operation U(op : (Qubit) => Unit is Adj + Ctl, power : Int,
    qubits : Qubit[]) : Unit is Adj + Ctl {
    for _ in 1 .. power {
        op(qubits[0]);
    }
}
```

Listing 7.22 A simple single qubit QPE oracle applying the desired power of U.

Notice that we defined the oracle with the controlled specialization so that it can be used as a controlled operation.

Unsurprisingly, Q# already provides a built-in implementation of quantum phase estimation algorithm, as the QuantumPhaseEstimation (oracle :   DiscreteOracle, targetState : Qubit[], control Register : BigEndian) operation from the Microsoft.Quantum. Characterization namespace. It is only available for the big endian register, but the qubit order can always be reversed into the little endian format if needed. The built-in functionality relies on the DiscreteOracle user defined type from Microsoft.Quantum.Oracles namespace, which is a wrapper for (Int, Qubit[]) => Unit is Adj + Ctl) delegate. For this reason, in order to be compatible with the DiscreteOracle wrapper user defined type, the signature of our oracle from Listing 7.22 takes an array of qubits as input, even though it only needs to act on one of them.

Notice, however, that the **U** operation defined by us is still not directly usable with the DiscreteOracle because of the extra op delegate parameter. We will be using the partial application feature with it, to prepare the U oracle with various single qubit operations which will serve as testing candidates. The oracle preparation function, to be used with partial application, is shown in Listing 7.23.

```
function PrepareOracle(op : (Qubit) => Unit is Adj + Ctl) :
    ((Int, Qubit[]) => Unit is Adj + Ctl) {
    return U(op, _, _);
}
```

Listing 7.23 Preparing the oracle Listing 7.22 with any arbitrary single qubit operation using Q# function application feature.

The output from the PrepareOracle (op : (Qubit) => Unit is Adj + Ctl) is a delegate that can be passed into DiscreteOracle. From this point on, we can take two possible paths—and in fact we will take both of them, comparing the outcomes. We could either rely on the built-in core quantum phase estimation functionalities of the QDK or we could implement the circuit Fig. 7.17 manually. Let us first begin with the manual procedure, which is shown in Listing 7.24.

```
operation ManualEstimation(
    eigenstate : Qubit,
    oracle : ((Int, Qubit[]) => Unit is Adj + Ctl),
    precision : Int) : Double {

    use qubits = Qubit[precision];
    let register = LittleEndian(qubits);

    ApplyToEach(H, qubits);

    for i in 0 .. precision - 1 {
        Controlled oracle([qubits[i]], (2^i, [eigenstate]));
    }

    Adjoint QFTLE(register);

    let phase = IntAsDouble(MeasureInteger(register)) /
        IntAsDouble(2^precision);
    return phase;
}
```

Listing 7.24 Manual implementation of the QPE from circuit Fig. 7.17.

The operation ManualEstimation (eigenstate: Qubit, oracle: ((Int, Qubit[]) => Unit is Adj + Ctl), precision : Int) takes a qubit in an eigenstate as an input (recall that we are using single qubit gates in this example, so a single qubit eigenstate is appropriate), the oracle delegate and the precision integer, which will dictate the size of the result register, as well as the amount of iterations (and thus powers) of application of the controlled oracle. The result register is allocated and placed into a superposition, followed by the sequence of controlled oracle calls, determined by the required precision, using the Controlled functor.

Next, an adjoint QFTLE (qs : LittleEndian operation from the Microsoft.Quantum.Canon namespace is executed. While we do want to focus on manual QPE algorithm implementation here, we already covered manual QFT in Sect. 7.4 so there is no need to repeat it again.

Finally, the phase is obtained by measuring the register in the little endian format. It will contain the phase value in radians, and depending on the precision, so, really, the amount of controlled operations invoked, we need to divide it accordingly.

The QDK facilitated QPE is shown in Listing 7.25. The main difference is that the built-in quantum phase estimation algorithm produces the result in the big endian register so prior to measurement with the MeasureInteger(target : LittleEndian) we need to manually reverse the qubits to be compatible with little endian bit order.

```
operation BuiltinEstimation(
    eigenstate : Qubit,
```

```
    oracle : ((Int, Qubit[]) => Unit is Adj + Ctl),
    precision: Int) : Double {

    use qubits = Qubit[precision];
    QuantumPhaseEstimation(DiscreteOracle(oracle), [eigenstate],
        BigEndian(qubits));
    let phase = IntAsDouble(MeasureInteger(LittleEndian(Reversed(
        qubits)))) / IntAsDouble(2^precision);
    return phase;
}
```

Listing 7.25 Implementation of the algorithm in Q# using the built-in QPE feature.

Both of the solutions should produce the exact same output, and it should be fairly easy to verify. Good candidates for testing are quantum gates **Z**, **S** and **T**. As we covered already in Sect. 4.5, they have a special relationship, namely $\mathbf{T}^4 = \mathbf{S}^2 = \mathbf{Z}$. Additionally they all add a phase to the quantum state $|1\rangle$:

$$\mathbf{Z}|1\rangle = -|1\rangle = e^{i\pi}|1\rangle \tag{7.106}$$

$$\mathbf{S}|1\rangle = i|1\rangle = e^{\frac{\pi i}{2}}|1\rangle$$
$$\mathbf{T}|1\rangle = e^{\frac{\pi i}{4}}|1\rangle$$

Because QPE allows us to estimate the phase in the form of $e^{2\pi i\theta}$, we should be able to find:

$$\theta_Z = \frac{1}{2} \quad \theta_S = \frac{1}{4} \quad \theta_T = \frac{1}{8} \tag{7.107}$$

To be able to compare the manual implementation and the built-in one, we define one other helper operation, that will take the prepared oracle and the relevant eigenstate as input and execute both flows. The output of this orchestrator can then be used to find out whether we made any mistakes in our theoretical reasoning. Since the biggest value we need to find, 8, requires three bit representation, we can set 3 as the precision level.

```
operation TestPhaseEstimation(
    oracle : (Int, Qubit[]) => Unit is Adj + Ctl,
    eigenstate : Qubit,
    expectedPhase : Double) : Unit {

    let libPhase = BuiltinEstimation(eigenstate, oracle, 3);
    let manualPhase = ManualEstimation(eigenstate, oracle, 3);

    Message($"Expected: {expectedPhase}");
    Message($"Library: {libPhase}");
    Message($"Manual: {manualPhase}");
    Message("");

    Reset(eigenstate);
}
```

Listing 7.26 Orchestrating both variants of QPE implementation for comparison purposes.

Finally, we need to invoke the orchestrator operation for **Z**, **S** and **T**. This will also be the entry point to our program.

```
@EntryPoint()
operation Main() : Unit {
    use eigenstate = Qubit();

    TestPhaseEstimation(
        PrepareOracle(Z),
        PrepareEigenState(eigenstate), 0.5);
    TestPhaseEstimation(
        PrepareOracle(S),
        PrepareEigenState(eigenstate), 0.25);
    TestPhaseEstimation(
        PrepareOracle(T),
        PrepareEigenState(eigenstate), 0.125);
}

operation PrepareEigenState(eigenstate : Qubit) : Qubit {
    X(eigenstate);
    return eigenstate;
}
```

Listing 7.27 Testing QPE for **Z**, **S** and **T** using the orchestrator from Listing 7.26.

Running it should confirm our expectations regarding the values of θ_Z, θ_S and θ_T.

```
Expected: 0.5
Library: 0.5
Manual: 0.5

Expected: 0.25
Library: 0.25
Manual: 0.25

Expected: 0.125
Library: 0.125
Manual: 0.125
```

Listing 7.28 Output of the code from Listing 7.27.

7.6 Shor's Algorithm

Shor's algorithm was first developed by Peter Shor in 1994 [23], and further refined by him in 1997 [24]. It is to this day the most spectacular example of a computational advantage that can be unlocked through quantum computing. It allows factoring integers with exponential speed-up compared to the best known classical alternative, the general number field sieve algorithm. The algorithm consists of classical pre-processing, a quantum core and classical post-processing.

Thanks to number theory we know that we can reduce the problem of factoring the product of two prime numbers to the problem of finding the period of a related *periodic function.* Based on this, Shor utilized the procedure of Quantum Fourier Transform, and, more generally, quantum phase estimation, which lies at the heart of his algorithm, to perform a very efficient period estimation. It suffices to say that Shor's proposal sparked a massive interest in the field of quantum computing, and provided a research momentum that is still noticeable even today, almost three decades later.

Shor's algorithm was implemented on a physical quantum hardware for the first time by Chuang et al. in 2001, who successfully factored the number 15 [26] using seven spin-$\frac{1}{2}$ nuclei in a molecule as quantum bits, manipulated through nuclear magnetic resonance. In 2012, factoring of the number 21 was achieved [20], and remains the highest number factored using the full Shor's algorithm on a physical quantum device. A simplified variant of Shor's algorithm, designed to work exclusively with composite numbers given by products of the Fermat primes, has been shown by Geller and Zhou [12] who demonstrated factoring of 51 and 85 using only 8 qubits. Improvements related to the amount of resources needed by Shor's algorithm include work by Beauregard [1], Zalka [27] and Gidney [13].

The reason why factorization of integers is a problem of such significance, is that we know[4] that due to the exponential nature of the involved complexity, it is an insurmountable challenge for classical computers. Because of that, factors of very large integers are used to underpin and to effectively guarantee the security of many cryptographic schemes used today, in particular the public-private RSA (Rivest–Shamir–Adleman) cryptosystem, which secures most of the data transmission on the Internet. It then follows, that using Shor's procedure, a sufficiently powerful quantum device, with appropriate error-correction and enough stability, could theoretically one day be used to break such cryptosystems that are relying on integer factorization—a particularly attractive proposition for certain (mostly evil) actors, and a dooming catastrophe for most of others.

Realistically speaking, we are far from breaking RSA with its 2048-bit (or larger) long keys. The current best estimation for breaking RSA integers, given the noise levels of today's quantum hardware, is provided by Gidney and Ekerå at 20 million qubits and 8h run time for the procedure [14]. Capacity-wise this of course far exceeds any quantum computer we can even imagine building in the near future, even though this estimation may go down by possible orders of magnitude as the error correction levels and fidelity improves. At the same time, it still illustrates the dramatic exponential speed up over classical methods, as it is roughly estimated that factoring RSA 2048 bit integer with a typical desktop computer, would take a 10^{15} years [4], which is longer than the age of the Universe.

An interesting approach suggested recently by Gouzien and Sangouard [15] involves the use of temporally and spatially multiplexed memory in the computer architecture such that qubit states can effectively be stored when they are unprocessed, which in turn dramatically reduces the number of compute qubits required as part of the QPU. With such approach, they estimated 2048-bit RSA integer to be factored in 177 days using a processor made up of 13436 physical qubits.

---

[4] Technically speaking there is no proof for that, though no efficient solutions have been found.

7.6.1 Period of a Function

Suppose we would like to factor number 15. The example is not particularly difficult, because it is obvious that

$$15 = 5 \times 3 \qquad (7.108)$$

however, because it is so basic, we should be able to follow it very easily. Let us consider the following function

$$f(x) = a^x \pmod{N} \qquad (7.109)$$

where a can take the values of $\{2, 3, \ldots, N - 1\}$ and is relatively prime to N. Two numbers are relatively prime[5] to each other when their greatest common divisor, which can be calculated using a classical algorithm, such as Euclid's algorithm, is equal to one.

$$gcd(a, N) = 1 \qquad (7.110)$$

Now, if we begin evaluating the function with $x = 0$ we obviously get

$$f(0) = a^0 \pmod{N} = 1 \qquad (7.111)$$

regardless of the value of a. The period r of this function, for a specific relatively prime pairs of a and N is then the smallest number $x > 0$, such that

$$f(n) = 1 \qquad (7.112)$$

We can use this knowledge to actually factor an $N = 15$. Suppose we start by setting $a = 13$ and carry on evaluating the values of the function $f(x)$ according to Eq. (7.109). This is illustrated on Fig. 7.18. We can quickly spot that

$$f(4) = 1 \qquad (7.113)$$

and a pattern of repetition emerged at the intervals of four, such that

$$f(4) = f(8) = f(12) = f(16) = 1 \qquad (7.114)$$

indicating that the period of this function is $r = 4$.

The next step is to check if the period r is an even or an odd number, as we can only continue when it is even, which is the case for our us, since $r = 4$. Additionally, we need to make sure that the following condition is not satisfied by our r

$$a^{\frac{r}{2}} = -1 \pmod{N} \qquad (7.115)$$

---

[5] They are also often referred to as *co-primes*.

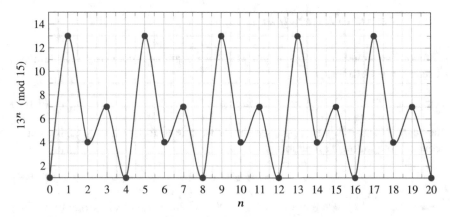

Fig. 7.18 Periodic function $f(n) = 13^n \pmod{15}$

Finally, number theory says that nontrivial factors of N will be two numbers p and q such that

$$p = gcd(a^{\frac{r}{2}} + 1, N) \tag{7.116}$$
$$q = gcd(a^{\frac{r}{2}} - 1, N)$$

In our case that will be

$$p = gcd(13^2 + 1, 15) = gcd(170, 15) = 5 \tag{7.117}$$
$$q = gcd(13^2 - 1, 15) = gcd(168, 15) = 3$$

And indeed, both 5 and 3 happen to be *non-trivial* factors of 15.

7.6.2 Quantum Order Finding

The circuit for the quantum part of the Shor's algorithm, allowing an efficient estimation of the period r, is shown in Fig. 7.19. If it looks familiar, it is not a coincidence—the quantum part of the Shor's algorithm is a de facto quantum phase estimation problem, and we already studied an almost identical circuit in Fig. 7.17. The important aspect here is that the bottom register is initialized to state $|0 \ldots 1\rangle$, and we will get to reasons behind that in a moment.

The top register, to which we will refer as the source register, needs to be of size $n = 2m$, where m corresponds to the bit size of the number being factored. For example, number 15 is a four bit integer, so $n = 8$. Just as discussed in the quantum phase estimation section earlier, its size will influence the precision of the solution. The bottom register, which we will refer to as the target register, will have the size m and will not be measured.

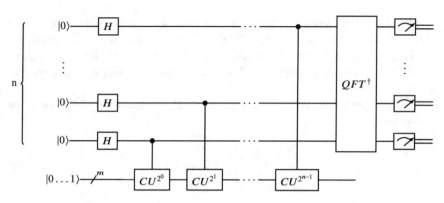

Fig. 7.19 Period estimation within Shor's algorithm

The execution of the routine on the quantum hardware requires us to be able to construct the unitary operator **U** such that

$$\mathbf{U}\,|x\rangle = \begin{cases} |ax \pmod{N}\rangle & \text{when } 1 \le x \le N-1 \\ |x\rangle & \text{otherwise} \end{cases} \tag{7.118}$$

In addition, we need to be able to use this operator in a way that supports modular exponentiation. Thankfully, as we learnt in Sect. 7.5, when applying the same operation multiple times, the resulting composite operation ends up being exponentiation of the original one.

$$\mathbf{U}^n\,|x\rangle = |a^n x \pmod{N}\rangle \tag{7.119}$$

As Fano and Blinder [11] point out, it may seem like a calculation of the period is an insurmountable challenge due to the enormous amount of permutations required for each of the candidate values within **U**. As it turns out, the quantum solution exploits the eigenstates $|u_j\rangle$ of the operator **U**, which are given by Nielsen and Chuang [22]

$$|u_j\rangle = \frac{1}{\sqrt{r}} \sum_{k=0}^{r-1} e^{-\frac{2\pi i jk}{r}} |a^k \pmod{N}\rangle \tag{7.120}$$

This can be tested by applying **U** to $|u_j\rangle$, which will lead us to the corresponding eigenvalue λ_j

$$\mathbf{U}\,|u_j\rangle = \mathbf{U}\left(\frac{1}{\sqrt{r}} \sum_{k=0}^{r-1} e^{-\frac{2\pi i jk}{r}} |a^{k+1} \pmod{N}\rangle\right) = e^{\frac{2\pi i j}{r}} |u_j\rangle \tag{7.121}$$

$$\lambda_j = e^{\frac{2\pi i j}{r}}$$

where $0 \le j \le r-1$.

At this point, we know we are able to use such eigenstate, and, by applying the phase estimation procedure from Eq. (7.105), to estimate the value of $\frac{j}{r}$. However, it seems that if we are able to prepare the eigenstate $|u_j\rangle$, we need to know the value of r in the first place, which seemingly defeats the purpose of executing the algorithm at all.

What can be exploited here, is that the sum of all the eigenstates of U satisfies the following

$$\frac{1}{\sqrt{r}} \sum_{j=0}^{r-1} |u_j\rangle = |1\rangle \tag{7.122}$$

namely, all the phase differences cancel out, leaving only $|1\rangle$, which of course is trivial to prepare as input into the algorithm and which is then used in the phase estimation procedure. We need to emphasize that it is indeed the state $|1\rangle$ encoded using m qubits e.g. $|000\ldots01\rangle$, not and entire register with every qubit set to $|1\rangle$, e.g. $|111\ldots11\rangle$

Since the outcome of the phase estimation produces the eigenvalue $\frac{j}{r}$ instead of r, it still need some post processing in order to extract the actual integer value of r. This can be done using the *continued fractions algorithm*, which allows us to explicitly extract values of j and r from the phase estimation result. For example, if the measured phase has the value of $0.333\ldots$, the algorithm can calculate $j = 1$ and $r = 3$.

Let us walk through the quantum steps with $N = 15$ and $a = 13$ as an example. Since we need 4 bits to express 15, we set $m = 4$ and $n = 8$ and start with the system state $|\psi_1\rangle$

$$|\psi_1\rangle = |00000000\rangle \otimes |0001\rangle \tag{7.123}$$

where the second register uses 4 qubits to encode $|1\rangle$. After the $\mathbf{H}^{\otimes 8}$ transformation is applied to the entire top register, we arrive at the usual superposition state

$$|\psi_2\rangle = \frac{1}{\sqrt{256}} \sum_{k=0}^{255} |k\rangle\, |0001\rangle \tag{7.124}$$

Next, we allow the $\mathbf{U}$ gate (or a series of its applications) to enact the transformation corresponding to the function $f(x) = a^x \pmod{N}$ leading us to

$$|\psi_3\rangle = \frac{1}{\sqrt{256}} \sum_{k=0}^{255} |k\rangle\, \big|13^k \pmod{15}\big\rangle \tag{7.125}$$

The state $|\psi_3\rangle$ can re-written into a more verbose form by expanding the sum into individual terms and by using the full number notation instead of binary representation inside the kets. This approach helps to highlight that we now deal with the entangled registers, and the periodic function really stands out clearly.

$$|\psi_3\rangle = \frac{1}{\sqrt{256}}\Big(|0\rangle\,|f(0)\rangle + |1\rangle\,|f(1)\rangle + \cdots + |255\rangle\,|f(255)\rangle\Big) = \quad (7.126)$$

$$\frac{1}{16}\Big(|0\rangle\,|1\rangle + |1\rangle\,|13\rangle + |2\rangle\,|4\rangle + |3\rangle\,|7\rangle + |4\rangle\,|1\rangle + |5\rangle\,|13\rangle + \cdots + |255\rangle\,|7\rangle\Big) =$$

$$\frac{1}{2}\Big(|0\rangle + |4\rangle + \cdots + |252\rangle\Big)\,|1\rangle + \frac{1}{2}\Big(|1\rangle + |5\rangle + \cdots + |253\rangle\Big)\,|13\rangle +$$

$$\frac{1}{2}\Big(|2\rangle + |6\rangle + \cdots + |254\rangle\Big)\,|4\rangle + \frac{1}{2}\Big(|3\rangle + |7\rangle + \cdots + |255\rangle\Big)\,|7\rangle$$

After applying the inverse Quantum Fourier Transform to this state we can measure, with equal probability distribution, one of four possible outcomes—$|0\rangle$, $|64\rangle$, $|128\rangle$ or $|192\rangle$. The corresponding phase is then the measurement result divided by $2^8 = 256$ giving us possible phases 0, 0.25, 0.5 and 0.75, which can then be used as input into continued fraction expansion. We can see from here that two of these values will not be usable. In the case of 0, we will calculate $r = 1$ which is invalid. On the other hand, 0.5 will expand to $\frac{1}{2}$ instead of $\frac{2}{4}$ because j and r are not coprimes, which in turns leads us to estimating a factor of r ($r = 2$) instead of proper period r value. The other two possible measurement outcomes will give us correct values of $r = 4$. Once we have the correct period, the quantum part of the algorithm ends.

7.6.3 Summary of Steps

Before we move to implementing the algorithm in Q#, let us now recap the **steps of the Shor's algorithm** to factor an integer N.

1. Pick a random a in the range of $1 < a < N$
2. Find the greatest common divisor c between a and N.

   - if $c \neq 1$ we have managed to guess the factor
   - if $c = 1$, we know that c and a are coprimes and the algorithm can continue

3. Using quantum logic, find the phase $\frac{j}{r}$ of a quantum version of $f(x) = a^x$ (mod N)
4. Using classical logic based on continued fraction expansion, estimate r from $\frac{j}{r}$

   - if r is odd or $a^{\frac{r}{2}} = -1$ (mod N), break the procedure and move on to a different a
   - otherwise, the algorithm can continue

5. Using classical logic, calculate the factor guesses p and q with Eq. (7.116), and check whether they are non-trivial e.g. they are larger than 1 and smaller than N

   - if p or q are non-trivial, we have found the factor
   - otherwise, break the procedure and move on to a different a

7.6.4 Q# Implementation

In order not to bloat the implementation logic of the algorithm in Q# we shall use hardcoded values of the number to be factored and the coprime used to begin the calculation with. We will set them, according to our earlier discussion to $a = 13$ and $N = 15$, as this will allow us to easily compare the program outcome with the theory. This approach can naturally be, with relatively modest effort, extended by the reader into a version that iterates over larger set of candidate a in the range of $1 < a < N$, and which supports dynamic values of N as program input.

The overall application is structurally crafted around three callables corresponding to different steps of Shor's algorithm and is shown in Listing 7.29. This procedure is executed repeatedly using the `repeat...until` Q# loop, until at least one non-trivial factor is found or, to cap the execution time of the program, a maximum of five attempts are processed. If any are discovered, the factors are then printed out and the program terminates.

```
@EntryPoint()
operation Main() : Unit {
    // N and a are hardcoded for simplicity
    let N = 15; // number to factor
    let a = 13; // coprime

    mutable factors = [];
    mutable attempt = 0;
    repeat {
        set attempt += 1;
        let estimatedPhase = EstimatePhase(a, N);
        let estimatedPeriod = CalculatePeriodFromPhase(
            estimatedPhase, N);
        set factors = CalculateFactors(a, N, estimatedPeriod);
    } until not IsEmpty(factors) or attempt >= 5;

    for factor in factors {
        Message($"Non trivial factor discovered: {factor}");
    }
}
```

Listing 7.29 Overall structure of the Shor program, according to the steps defined in Sect. 7.6.3.

The quantum logic is encapsulated in the operation `EstimatePhase` (a : Int, N : Int), which executes phase estimation and gives us back the value of $\frac{i}{r}$. Next, the phase is passed into a classical function `CalculatePeriodFromPhase` (phase : Int, N : Int) which utilizes continued fraction expansion to calculate the value of r. Finally, another classical function `CalculateFactors` (a : Int, N : Int, r : Int) validates the correctness of r and converts the estimated r into the actual p and q factor candidates.

The phase estimation logic is shown in Listing 7.30. It is based on the QPE that is already built into the Q# standard library and follows the approach that we covered in Listing 7.25. As we mentioned already, due to the fact that library based phase estimation operates on big endian qubit registers, we need to manually reverse the qubits to be compatible with our preferred little endian bit order. The register sizes are determined by the bit size of the number N. Because the target register is never measured, it needs to be reset at the end of the operation.

```
operation EstimatePhase(a : Int, N : Int) : Int {
    // size of the registers is determined by bitsize of number to factor
    let numToFactorBitSize = BitSizeI(N);
    if numToFactorBitSize > 8 {
        fail "Maximum supported integer size is 8 bit.";
    }

    // init target register to number 1
    use target = Qubit[numToFactorBitSize];
    X(target[numToFactorBitSize-1]);

    let oracle = DiscreteOracle(PrepareOracle(a, N, _, _));
    use source = Qubit[2 * numToFactorBitSize];

    // run quantum phase estimation
    QuantumPhaseEstimation(oracle, target, BigEndian(source));
    let phaseResult = MeasureInteger(
        LittleEndian(Reversed(source))
    );

    Message($"Phase: {IntAsDouble(phaseResult)
        / IntAsDouble(2^(2 * numToFactorBitSize))}");
    ResetAll(target);
    return phaseResult;
}
```

Listing 7.30 Phase estimation in Q#, as part of Shor's algorithm.

The transformation **U** is constructed around the `MultiplyByModularInteger` (`multiplier : Int, modulus : Int, multiplier : LittleEndian`) operation from the `Microsoft.Quantum.Arithmetic` namespace, which already implements the required unitary operation $|ax \pmod N\rangle$[6] and is shown in Listing 7.31. The exponentiation is ensured by running the multiplier through the `ExpModI (expBase : Int, power : Int, modulus : Int)` function from the `Microsoft.Quantum.Math` namespace. Because of how Q# core libraries define parameters on the built-in quantum phase estimation operation (the requirement to wrap the logic of **U** with the `DiscreteOracle` user-defined type), we need to use the partial callable application feature to ensure the signatures match appropriately.

```
operation PrepareOracle(a : Int, N : Int, pow : Int, target :
    Qubit[])
    : Unit is Adj + Ctl {
    let multiplier = ExpModI(a, pow, N);
    let register = LittleEndian(target);
    MultiplyByModularInteger(multiplier, N, register);
}
```

Listing 7.31 The Q# code for the $U^n |x\rangle = |a^n x \pmod N\rangle$ transformation as part of Shor's algorithm.

---

[6] I am indebted to Cassandra Granade for this solution, used in the initial implementation of Shor's algorithm in QDK samples [16].

Next the continued fraction expansion is used to convert the obtained measurement result into $\frac{i}{r}$ integer-based representation, as shown in Listing 7.32. Yet again, to our convenience, the continued fraction algorithm is built into the core Q# libraries, as ContinuedFractionConvergentI (fraction : Fraction, denomBound : Int) function from the Microsoft.Quantum.Math namespace, allowing us to determine the denominator r from the fraction $\frac{i}{r}$.

```
function CalculatePeriodFromPhase(measurementResult : Int, N :
    Int) : Int {
    // convert value like 0.75 to 3/4
    let input = Fraction(measurementResult, 2^(2 * BitSizeI(N)));
    let phase = ContinuedFractionConvergentI(input, N);
    Message($"Estimated value for r is
        {AbsI(phase::Denominator)}");
    return AbsI(phase::Denominator);
}
```

Listing 7.32 Continued fraction expansion in Q# code, as part of Shor's algorithm.

The final building block is a dedicated function shown in Listing 7.33, which determines whether the estimated r needs to be discarded and a subsequent attempt is needed. This happens when r is 0, r is odd or $a^{\frac{r}{2}} = -1 \pmod{N}$. When none of these conditions is satisfied, the factor candidates p and q are calculated using Eq. (7.116) and verified that they are non-trivial.

```
function CalculateFactors(a : Int, N : Int, r : Int) : Int[] {
    if r == 0 {
        Message("r is 0");
        return [];
    }

    if r % 2 != 0 {
        Message("r is odd");
        return [];
    }

    let aToHalfR = a ^ (r / 2);
    if aToHalfR == N - 1 {
        Message($"{a}^({r}/2) = -1 mod {N}");
        return [];
    }

    // period is valid, find factor candidates based on gcd
    let candidates = [
        GreatestCommonDivisorI(aToHalfR + 1, N), // p
        GreatestCommonDivisorI(aToHalfR - 1, N)]; // q

    // filter out trivial factors
    let filter = IsNonTrivialFactor(N, _);
    return Filtered(filter, candidates);
}
```

```
function IsNonTrivialFactor(N : Int, i : Int) : Bool {
    return i > 1 and i < N and N % i == 0;
}
```

Listing 7.33 Q# code to calculate factors based on the estimate r, as part of Shor's algorithm.

At this point we should be to test the program and verify its output. Running the program several times can confirm that, and should produce the output similar to that in Listing 7.34. If we are unlucky we may see the invalid value of phase 0, and a retry attempt that followed it.

```
Phase: 0.75
Estimated value for r is 4
Non trivial factor discovered: 5
Non trivial factor discovered: 3
```

Listing 7.34 Sample output of the Q# implementation of the Shor algorithm when factoring 15.

References

1. Beauregard, S. (2003). Circuit for Shor's algorithm using 2n+3 qubits.
2. Bera, R. K. (2020). *The amazing world of quantum computing*. Singapore: Springer.
3. Bernstein, E., & Vazirani, U. (1997). Quantum complexity theory. *SIAM Journal on Computing, 26*(5), 1411–1473.
4. Chen, L. (2017). Deciphering: The thrill of a lifetime. https://www.nist.gov/blogs/taking-measure/deciphering-thrill-lifetime. Retrieved 18 Jan 2022.
5. Chuang, I., Gershenfeld, N., & Kubinec, M. (1998). Experimental implementation of fast quantum searching. *Physical Review Letters - PHYS REV LETT, 80*(04), 3408–3411.
6. Cleve, R., Ekert, A., Macchiavello, C., & Mosca, M. (1998). Quantum algorithms revisited. *Proceedings of the Royal Society of London. Series A: Mathematical, Physical and Engineering Sciences, 454*(1969), 339–354.
7. Cooley, J., & Tukey, J. (1965). An algorithm for the machine calculation of complex Fourier series. *Mathematics of Computation, 19*(90), 297–301.
8. Coppersmith, D. (2002). An approximate Fourier transform useful for quantum factoring. 02.
9. Deutsch, D. (1985). Quantum theory, the Church-Turing principle and the universal quantum computer. *Proceedings of the Royal Society of London. Series A, Mathematical and Physical Sciences, 400*(1818), 97–117.
10. Deutsch, D., & Jozsa, R. (1992). Rapid solution of problems by quantum computation. Technical report, GBR.
11. Fano, G., & Blinder, S. M. (2017). *Twenty-first century quantum mechanics: Hilbert space to quantum computers*. Switzerland: Springer Nature.
12. Geller, M. R., & Zhou, Z. (2013). Factoring 51 and 85 with 8 qubits.
13. Gidney, C. (2018). Factoring with n+2 clean qubits and n-1 dirty qubits.
14. Gidney, C., & Ekerå, M. (2021). How to factor 2048 bit RSA integers in 8 hours using 20 million noisy qubits. *Quantum, 5*(Apr), 433.
15. Gouzien, E., & Sangouard, N. (2021). Factoring 2048-bit RSA integers in 177 days with 13436 qubits and a multimode memory. *Physical Review Letters, 127*(14).
16. Granade, Cassandra, Microsoft, & Contributors. (2022). Microsoft quantum development kit samples. https://github.com/microsoft/Quantum. Retrieved 16 Jan 2022.
17. Grover, L. K. (1996). A fast quantum mechanical algorithm for database search. In *Annual ACM Symposium on Theory of Computing* (pp. 212–219). ACM.

18. Hidary, J. D. (2019). *Quantum computing: An applied approach*. Switzerland: Springer Nature.
19. Kitaev, A. Y. (1996). Quantum measurements and the Abelian stabilizer problem. *Electronic Colloquium on Computational Complexity, 3.*
20. Martin-Lopez, E., Laing, A., Lawson, T., Alvarez, R., Zhou, X.-Q., & O'Brien, J. L. (2012). Experimental realization of Shor's quantum factoring algorithm using qubit recycling. *Nature Photonics, 6*(11), 773–776.
21. Mermin, N. D. (2007). *Quantum computer science. An introduction*. Cambridge: Cambridge University Press.
22. Nielsen, M. A., & Chuang, I. (2010). *Quantum computation and quantum information* (10th Anniversary ed.). Cambridge: Cambridge University Press.
23. Shor, P. W. (1994). Algorithms for quantum computation: Discrete logarithms and factoring. In *Proceedings of the 35th Annual Symposium on Foundations of Computer Science* (pp. 124–134).
24. Shor, P. W. (1997). Polynomial-time algorithms for prime factorization and discrete logarithms on a quantum computer. *SIAM Journal on Computing, 26*(5), 1484–1509.
25. Svore, K. M., Hastings, M. B., & Freedman, M. (2013). Faster phase estimation.
26. Vandersypen, L. M. K., Steffen, M., Breyta, G., Yannoni, C. S., Sherwood, M. H., & Chuang, I. L. (2001). Experimental realization of Shor's quantum factoring algorithm using nuclear magnetic resonance. *Nature, 414*(6866), 883–887.
27. Zalka, C. (2006). Shor's algorithm with fewer (pure) qubits.
28. Zygelman, B. (2018). *A first introduction to quantum computing and information*. Switzerland: Springer Nature.

Chapter 8
Where to Go Next?

We have managed to reach the end of this book, which means that this particular journey is inevitably nearing its conclusion. At the same time, and these words are accompanied by great optimism, hopefully for the reader this ending only marks a beginning of a fascinating and rewarding adventure with quantum technologies. There are quite a few possible ways to go further from here, and the present author will humbly allow himself to suggest several possible options.

The core field of our interest in this study, quantum computing, enjoys a wide spectrum of superb texts, so there are countless excellent follow-up reading propositions. One of the great minds of quantum information science, David Mermin, provides a very light-hearted, yet comprehensive, tour through quantum computing concepts [14], which is a must read for anyone studying this topic. Yanofsky and Mannucci [25] target their unique book specifically at computer scientists and cover quantum computing from that perspective. Rieffel and Polak [20] dive deeply into the principles of quantum computing, and cover more advanced topics such as error correction, fault tolerance or decoherence, in a very digestible way. Zygelman [27] does a spectacularly efficient job of going from the most basic concepts to design of quantum hardware in about 200 pages, and on his way includes countless links to Wolfram Mathematica® notebooks that illustrate the theory covered.

For the readers willing to continue more in the direction of the field of quantum information theory, a magnificent further reading could be Schumacher and Westmoreland [23], who do an excellent job of covering both typical aspects of quantum theory in a novel, yet structured way, as well as in linking those to concepts of quantum information science. The textbook of John Watrous [24], based on his lectures, is already a classic in teaching the theory of quantum information.

For a wide range of more advanced texts covering topics such as quantum machine learning [22], quantum error correction [11], quantum cryptography [7] or quantum search algorithms [19], to only name a few, the Quantum Science and Technology

© The Author(s), under exclusive license to Springer Nature Switzerland AG 2022
F. Wojcieszyn, *Introduction to Quantum Computing with Q# and QDK*, Quantum Science and Technology, https://doi.org/10.1007/978-3-030-99379-5_8

series from Springer is an indispensable source of valuable information, presented with great attention to detail.

And of course, it is impossible not to mention the phenomenal "Mike and Ike" textbook by Nielsen and Chuang [15], one of the great texts in physics history. It is an advanced reading, that requires full cognitive attention and can get overwhelming for beginners. At the same time, the textbook is delightful to follow and lends itself well to being enjoyed at one's preferred pace. It also is incredibly detailed, and, despite its age, is still a de facto complete compendium of quantum computing.

For readers interested in diving deeper into the Q# world, there is no better resource than the official Azure Quantum and Q# documentation.[1] Contrary to many other technical resources online, it is extremely well maintained, and contains a lot more in-depth information about the runtime and the language that we had a chance to cover in this book. It is also the recommended place to keep an eye on the latest developments in Azure Quantum and the changes related to both the software and the hardware make up of the platform.

The official Microsoft QDK samples repository on GitHub [6] is full of unique and valuable examples of solving specific quantum problems with the QDK. Finally, Granade and Kaiser [8] have an excellent Q#-specific book, approaching quantum computing from an original and entertaining angle, with a caring and thoughtful style.

We only minimally touched upon the core principles of quantum mechanics in this book. Readers enthusiastic about pursing this path further might find the textbook by Kok [9] to be an attractive proposition, as, using his tremendous teaching skills, he manages a comprehensive introduction into quantum mechanics using only basic algebra. A similar approach is the theme of the textbook by Zubairy [26], who does a fantastic job of exhaustive discussion on basics of quantum mechanics, based on high-school level mathematics only. Fano and Blinder [4] provide an excellent account of quantum mechanics in an end-to-end fashion, with special focus on concepts relevant to quantum computing. For a more advanced presentation a published lecture series by Basdevant [1] is an excellent choice.

Finally, the history and development of quantum mechanics, as well as its philosophical consequences are another formidably interesting space. An excellent account of the early development of quantum mechanics, with narrative that reads like a thriller, is given by Kumar [10]. The most definitive work chronicling the conceptual creation of quantum theory is the six volume series by Mehra and Rechenberg [13], beaming with detailed information, and spanning over 5000 (!) pages.

Friebe et al. [5] is arguably the most well-rounded, structured introduction into philosophy of quantum physics. Laloë [12] uses a very unique mix of mathematics and philosophy, when he poses a question whether we really understand quantum mechanics. Plotnitsky, whose eloquent writing style is second to none, covers the nature of quantum-theoretical thinking [17], tackles the epistemological and ontological nature of reality [18] and attempts to decipher Bohr's texts [16]. In his editorial

---

[1] https://docs.microsoft.com/en-us/azure/quantum/.

contribution to The Frontiers Collection series of Springer, Schlosshauer [21] talks to some of the greatest physicists of recent times about a rich collection of topics such as the measurement problem or the consequences of Bell's inequalities, obtaining, as expected, an entertainingly wide spectrum of answers, anecdotes and opinions. A similarly absorbing set of quantum theoretical and philosophical debates by a group of esteemed French physicists can be found in another title from the same series, compiled by the editors Zwirn and d'Espagnat [3]. Finally, readers interested in the consequences of Bell's theorem would undoubtedly be delighted with another publication from The Frontiers Collection series, a fascinating compendium of essays published to commemorate 50 years since Bell's discovery, edited by Bertlmann and Zeilinger [2].

References

1. Basdevant, J.-L. (2016). *Lectures on quantum mechanics*. Graduate texts in physics. Cham: Springer.
2. Bertlmann, R., & Zeilinger, A. (2017). *Quantum [un]speakables II*. The Frontiers collection. Cham: Springer.
3. d'Espagnat, B., & Zwirn, H. (2016). *The quantum world*. The Frontiers collection. Cham: Springer.
4. Fano, G., & Blinder, S. M. (2017). *Twenty-first century quantum mechanics: Hilbert space to quantum computers*. Switzerland: Springer Nature.
5. Friebe, C., Kuhlmann, M., Lyre, H., Näger, P. M., Passon, O., & Stöckler, M. (2018). *The philosophy of quantum physics*. New York: Springer International Publishing AG.
6. Granade, Cassandra, Microsoft, & Contributors. (2022). Microsoft quantum development kit samples. https://github.com/microsoft/Quantum. Retrieved 16 Jan 2022.
7. Grasselli, F. (2021). *Quantum cryptography*. Quantum science and technology. Cham: Springer.
8. Kaiser, S. C., & Granade, C. (2021). *Learn quantum computing with Python and Q#: A hands-on approach*. New York: Manning.
9. Kok, P. (2018). *A first introduction to quantum physics*. New York: Springer International Publishing.
10. Kumar, M. (2009). *Quantum*. London: Icon Books Ltd.
11. La Guardia, G. G. (2020). *Quantum error correction*. Quantum science and technology. Cham: Springer.
12. Laloë, F. (2019). *Do we really understand quantum mechanics?* Cambridge: Cambridge University Press.
13. Mehra, J., & Rechenberg, H. (2001). *The historical development of quantum theory 1–6*. New York: Springer.
14. Mermin, N. D. (2007). *Quantum computer science. An introduction*. Cambridge: Cambridge University Press.
15. Nielsen, M. A., & Chuang, I. (2010). *Quantum computation and quantum information* (10th Anniversary ed.). Cambridge: Cambridge University Press.
16. Plotnitsky, A. (2006). *Reading Bohr: Physics and philosophy*. Netherlands: Springer.
17. Plotnitsky, A. (2010). *Epistemology and probability: Bohr, Heisenberg, Schrödinger, and the nature of quantum-theoretical thinking*. New York: Springer.
18. Plotnitsky, A. (2016). *The principles of quantum theory, from Planck's Quanta to the Higgs Boson*. New York: Springer International Publishing.
19. Portugal, R. (2018). *Quantum walks and search algorithms*. Quantum science and technology. Cham: Springer.

20. Rieffel, E. G., & Polak, W. H. (2014). *Quantum computing: A gentle introduction*. Cambridge: MIT Press.
21. Schlosshauer, M. (Ed.). (2011). *Elegance and Enigma*. The Frontiers collection. Berlin: Springer.
22. Schuld, M., & Petruccione, F. (2021). *Machine learning with quantum computers*. Quantum science and technology. Cham: Springer.
23. Schumacher, B., & Westmoreland, M. (2012). *Quantum processes systems, and information*. Cambridge: Cambridge University Press.
24. Watrous, J. (2018). *The theory of quantum information*. Cambridge: Cambridge University Press.
25. Yanofsky, N. S., & Mannucci, M. A. (2008). *Quantum computing for computer scientists*. Cambridge: Cambridge University Press.
26. Zubairy, M. S. (2020). *Quantum mechanics for beginners*. New York: Springer International Publishing AG.
27. Zygelman, B. (2018). *A first introduction to quantum computing and information*. Switzerland: Springer Nature.

Index

© The Editor(s) (if applicable) and The Author(s), under exclusive license
to Springer Nature Switzerland AG 2022
F. Wojcieszyn, *Introduction to Quantum Computing with Q# and QDK*, Quantum Science
and Technology, https://doi.org/10.1007/978-3-030-99379-5

Printed in the United States
by Baker & Taylor Publisher Services